S0-CAH-652

The Development of Modern Biology

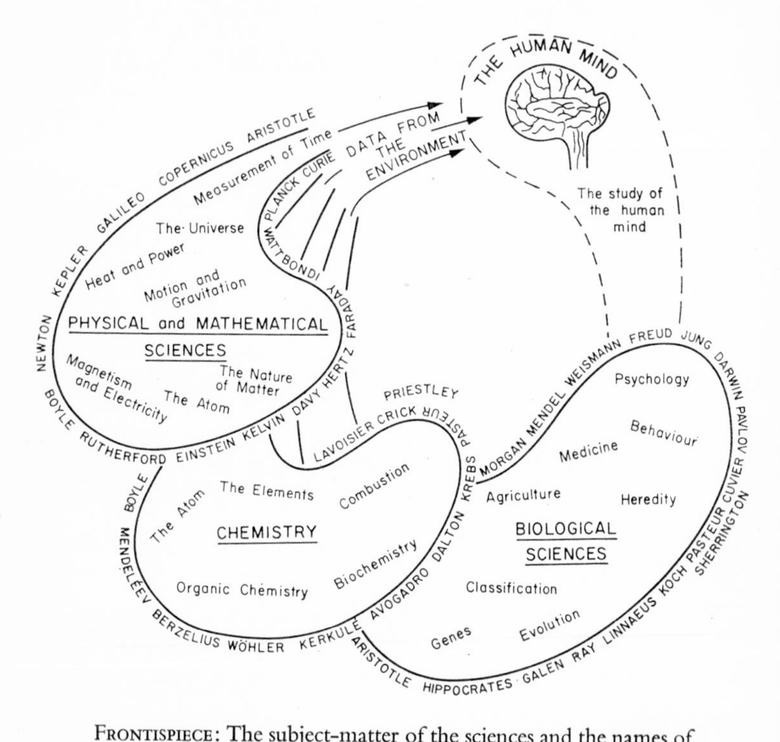

FRONTISPIECE: The subject-matter of the sciences and the names of the great contributors to our scientific knowledge. The human mind, which analyses the data of its environment, has itself, in recent years, become the subject of scientific analysis and study.

The Development of Modern Biology

A TEXT FOR HISTORY OF SCIENCE EXAMINATIONS AND BACKGROUND READING IN BIOLOGICAL SCIENCE FOR VI FORMS

by

P. T. MARSHALL, M.A.

Senior Biology Master at The Leys School, Cambridge and Awarder in History of Science to the Oxford and Cambridge Schools Examination Board

PERGAMON PRESS

OXFORD · LONDON · EDINBURGH · NEW YORK
TORONTO · SYDNEY · PARIS · BRAUNSCHWEIG

Pergamon Press Ltd., Headington Hill Hall, Oxford
4 & 5 Fitzroy Square, London W.1
Pergamon Press (Scotland) Ltd., 2 & 3 Teviot Place, Edinburgh 1
Pergamon Press Inc., Maxwell House, Fairview Park, Elmsford, New York 10523
Pergamon of Canada Ltd., 207 Queen's Quay West, Toronto 1
Pergamon Press (Aust.) Pty. Ltd., 19a Boundary Street, Rushcutters Bay, N.S.W. 2011, Australia
Pergamon Press S.A.R.L., 24 rue des Écoles, Paris 5e
Vieweg & Sohn GmbH, Burgplatz 1, Braunschweig

First edition 1969

Library of Congress Catalog Card No. 70-92112

Printed in Great Britain by The Cavendish Press Limited, Leicester.

08 006640 2 (flexicover)
08 013259 6 (hard cover)

Contents

List of Plates

Acknowledgements

THE author would like to acknowledge the debt he owes to the lectures on the history of biology delivered by the late Canon Raven in the University of Cambridge in 1953.

Figure 2.1 is redrawn from Knowles, *Man and Other Living Things*, Harrap. Figure 3.2 was derived from the British Museum publication *Evolution*. The scheme for Darwin's finches (Figure 3.1) was adapted from the Marian Ray film strip on Darwin. Figure 4.3 is redrawn from Wilma George, *Elementary Genetics*, Macmillan. Permission to use this material is acknowledged.

Thanks are also due to the following who have read and commented on various parts of the manuscript and who have made numerous helpful suggestions. Chapter 1, on the Development of Scientific Method, G. Buchdahl of the Department of History of Science, Cambridge University, and Dr. Brock of the Department of History of Science in the University of Leicester. Chapter 2, on Classification, Dr. John Gilmour, Director of the Botanic Garden, Cambridge University. Chapter 4, advice on genetics obtained from the Librarian of the Genetics Department of Cambridge University. Chapters 5 and 6, on Medicine, Dr. Gresham of the Pathology Department, Cambridge University. Chapter 7, on Agriculture, Dr. D. B. Wallace of the Department of Agriculture, Cambridge University, and Mr. Derek Baker and Mr. Martin Bell, colleagues at the Leys School.

Finally, to the Master and Fellows of Caius College, Cambridge, for permission to reproduce Plates I–IV taken from material in the college library, and to the Directors of the Plant Breeding Institute, Trumpington, for the abstract of current research projects.

CHAPTER 1

The Origins and Development of the Scientific Method

ENORMOUS material benefits have been brought about by the applications of scientific discoveries, and we may look on the development of science as one of the major advances of the human race.

It is possible to trace modern science back to a change in outlook and method which began in western Europe some 300 years ago. One of the main reasons for the success of this "new" science was due to the introduction of mechanical and mathematical analysis of phenomena and the use of experimentation as a means of gaining knowledge.

In this first chapter the ideas and methods of science and their origins will be described in outline so that the contents of the following, and specifically biological chapters, may be understood in the general context of scientific history.

Early Science

The civilizations of the Babylonians and the Egyptians flourished some 4000 years ago, and it is with them that the history of science begins. The notion of quantity and its exact measurement is implicit in all scientific procedures. In these early civilizations the study of mathematics and the division of weights, lengths, and times into units had begun; indeed, it seems that the Babylonians had even taken conceptual mathematics as far as the solving of quadratic equations.

Greek civilization greatly advanced the knowledge of science, and it was of them that John Stuart Mill said: "they were the most remarkable people that ever existed . . . they were the beginners of nearly everything that the modern world makes its boast." The Greeks saw "science" as a body of knowledge and the operation of the world as conforming to a universal and uniform law. To Plato and Aristotle there existed an order and pattern to events and phenomena that could be arrived at by thought and by intuition into the causes and principles on which they depended. This essentially reasonable attitude divorced such happenings as eclipses, tides and storms, disease, and the motion of the planets from the wills of erratic and uncertain Gods and is, of course, the basis of modern scientific philosophy.

Partly perhaps because of their "slave-powered" labour source, which gave little stimulus to the application of scientific innovation to technological problems, their contributions were mainly of an academic nature. However, in their belief that the world was capable of being understood by human intellect and that effects had preceding causes, they were the true forerunners of modern science.

It is by no means clear why the scientific revolution that took place in western Europe in the sixteenth and seventeenth centuries did not occur in the fairly settled and prosperous conditions of the great Roman Empire that followed the Greeks. Certainly, compared with the massive strides taken by the Greeks, the Romans themselves produced relatively few advances in science, although outstanding was the work of Ptolemy in astronomy and Galen in anatomy and physiology. They were by nature less original and philosophical, and much of their "science" was encylopedic and derived from their predecessors (see reference to Pliny, p 15). From the technological point of view the Romans, albeit another slave-powered civilization, were in advance of the Greeks.

With the break up of the Roman Empire, Europe entered a period of poverty and disorder which was to last nearly 500 years. This time has been referred to as the Dark Ages, but the term is somewhat misleading as at some times and in some places conditions were far from dark. In general, however, the lamp of knowledge burned low in the West, and such scholarship as existed was to be found in the monasteries.

Meanwhile into the hands of the triumphantly expanding Islam passed the scientific knowledge and traditions of the ancients. For the next five centuries the science of the Arabs was far in advance of that of the West, and their mathematics, astronomy, and medicine flourished.

As far as western Europe was concerned the monks were naturally more interested in theological rather than scientific problems. Typical of the times was the idea developed by St. Augustine in the fourth century that the whole earth had been created entirely for the benefit of man and that everything in it had some sort of spiritual significance. Such an anthropocentric point of view led to a curious and jaundiced interpretation of nature.

Of course some fragments of the old knowledge remained, and Boethius (480–524) translated some of Aristotle's philosophical writings as well as producing some derivative works on arithmetic. Likewise Bede (673–735) wrote a commentary on the Roman author Pliny.

It appears that by the end of the eleventh century the state of Europe had become more settled, and with the introduction of new agricultural methods, and in particular the heavier plough, there was a rise in population and prosperity. Christian armies began the re-conquest of Spain from the Moslems, and the Crusaders won back Sicily and began their excursions into the Near East. Partly through this warfare, and partly by trade and the movement of population, there arose a more intimate contact between the cultures of Islam and Europe. While rejecting the religion of the former, European scholars found they had much to learn from the knowledge of the Moslems. It was during this time (*c*. 1150) that some of the writings of Hippocrates, Aristotle, and Galen were translated from Arabic into Latin and became available in Europe.

From the twelfth and the fourteenth centuries the Christian schoolmen studied the ancient learning within the framework of the Church's teaching, their works being known as scholasticism. On the whole they believed that while man could give a description of the material world, only the authority of the Church could interpret this description. A critical commentary on the writings of Aristotle by Averroes (1126–98) was studied with great interest by these schoolmen in the new universities of Europe and in particular in those of Paris and Padua. Generally they believed the Greek interpretation to be correct, and we find Albertus Magnus (1206–80) producing writings based almost entirely on Aristotle. The one-time pupil of Magnus was St. Thomas Aquinas, and he was to produce a synthesis of the Church's teaching and Aristotle's philosophy (and both were based on design and purpose in nature) which became accepted as orthodoxy. Aquinas believed in the use of reason in the interpretation of phenomena and observation, but that as reason was a God-given gift it could never arrive at a conclusion different to the revealed knowledge of the Church. It became heretical to question these views and, indeed, it was in 1251 that the notorious Inquisition was set up by Pope Innocent III for the specific purpose of strengthening the disciplinary hold of the Catholic Church.

Despite such a check to freedom of thought in Western Christendom, various critical movements continued to exist throughout the thirteenth and fourteenth centuries. Men such as John Duns Scotus and William of Ockham began to question some of Aristotle's ideas of motion. On the Continent groups influenced by Averroes (whose teachings were regarded as heretical) argued for the principles of determinism whereby all things, including God, must have predetermined causes. Such men were against the rigid framework of the schoolmen, and in their various ways they began the undermining of the authority of the ancient learning.

As far as methods are concerned, we find Grosseteste and his more famous pupil Roger Bacon (1214–94) searching for new techniques of gaining knowledge. The one that they devised was the setting up of a number of hypotheses to explain a particular phenomenon and then, by a process of experimentation and trial and error, to eliminate false hypotheses and thus arrive at a probable explanation. This procedure is one of the most efficient methods of modern science, but although Grosseteste and Bacon did do some work on optics it is true to say that they wrote and thought more about their technique from a philosophical rather than practical point of view.

In retrospect we can see that whatever the limitations of scholasticism may have been, its adherents were at least responsible for a revival of learning in the West, and to them we may credit the "little scientific revolution" of the times. This little revolution was to lead to the much greater scientific revolution of succeeding centuries.

The Scientific Revolution

By the end of the fifteenth century the European sciences had more or less reached the limits of their Greek predecessors and were poised for the major advances that took place at the Renaissance. Some of the factors behind this latter period are discussed on p. 7, and it will be sufficient to say here that Europe was going through a period of intellectual, social, and material turmoil. In the first place the geographical boundaries of the known world had been greatly extended. (Columbus's discovery of the New World; the finding of a sea route to India via the Cape of Good Hope; and Magellan's circumnavigation of the globe in 1519.) From a purely utilitarian point of view it was necessary to improve navigational instruments, and this in itself demanded a reassessment of the cosmological views of the ancients.

There began to be a shift away from the contemplation of theological problems to observation of the external world, and in the field of art the use of perspective as a truer means of depicting objects was introduced. This realism in art led to a closer examination of the human body, and in the great Leonardo da Vinci (1452–1519) we find artist, anatomist, scientist, and engineer combined. Following the same pattern Versalius produced in 1543 his work on the internal anatomy of man which was based on actual dissections and clearly revealed many of the errors of Galen. In this same year Copernicus came out with his theory of a heliocentric universe with the earth and planets revolving round the sun, rather, as was then thought, that the sun moved around the earth. This theory was to find further overwhelming confirmation in the work of Galileo 70 years later.

The spread of these new ideas was assisted by the invention of printing and the use of woodcuts for making illustrations. Without the latter it would not have been possible for Versalius to have produced such a splendid book of anatomy.

In England a contemporary of Galileo, William Harvey (1578–1657) successfully questioned the teaching of Galen on the blood and its motion, and was able to show that the heart was in the nature of a mechanical pump which caused the blood to circulate. Galen himself had carried out some experiments on the motion of the blood, but he had not interpreted them correctly and had thought the blood to ebb and flow and to pass from one side of the heart directly to the other. The findings of Harvey lent support to the new mechanical interpretation of living organization and function although he himself did not subscribe to such views. (The culmination of his mechanistic philosophy is found in Descartes's writings.)

These and other advances were not established without considerable opposition from the orthodoxy left over from preceding centuries. Galileo was forced by the Inquisition to retract his ideas, and Versalius was never really prepared to refute Galen's now "invisible pores" in the septum. While William Harvey did not suffer for his theories, less fortunate was Servetus who was burnt by the Calvinists for his heretical views; these were mainly of a theological nature, but he also questioned Galen's views on circulation (a part of Church orthodoxy). Servetus mixed up his theological and scientific beliefs. Because of his views on circulation and the nature of the Trinity, he thought the soul to be in the blood, and this was one of the heresies of which he was accused. In fact the Reformers of the Church were by no means amicable to the advance of scientific ideas. Some of the early Protestants were, indeed, most hostile to Luther himself and ridiculed the ideas of Copernicus. (It is also true that well into the seventeenth century the puritans were very involved with various mystical ideas quite contrary to a mechanistic approach.)

Although a mention has already been made of the search for a technique of investigation, it must be appreciated that the promising methods of Grosseteste, Roger Bacon, and others had only partly come into use (as, for example, by William Harvey who attempted to back up all his suggestions by experimental proof). The actual development of a "scientific method" such as we would recognize it today took a long time to come about, and, perhaps more valuable than anything else, it was the outstanding contribution of the seventeenth century. One who advocated the use of experimentation was Francis Bacon, a lawyer and eventually Lord Chancellor under James I. Bacon was interested in science, and he said of Nature that "like a witness, she reveals her secrets when put to torture". Now while it is clear that hypothesis testing by experiment is an important part of modern scientific procedure, it is also clear that original hypotheses are often formed by intuition or by chance rather than by reason. In all great scientific advances there seem to be flashes of a genius and intuition that cannot be analysed into a circumscribed "method".

By 1600 there were a good number of men engaged, whether as amateurs or professionals, in the pursuit of science, and it is in this century we find the establishment of a number of scientific societies. In such gatherings, often patronized by rich men who had not time or talent for individual research but where interested in its findings, we get the mutual exchange of ideas so necessary for scientific progress. The change of science from the private pastime of a few eccentric individuals to being socially acceptable was indeed a further important advance.

The Accademia dei Lincei to which Galileo belonged was founded in Rome and was one of the first scientific societies. (It broke up at the refutation of Copernicus' work by the Church in 1630.) Shortly afterwards in northern Europe the Académie Royale des Sciences was started in Paris, and in 1662 the Royal Society of London received its charter from Charles II of England.

The founder members of the Royal Society included Robert Boyle, Robert Hooke (p. 83), and Willughby (p. 17) among others, and it was through this society that the work of Isaac Newton became known. His *Principia* is one of the great books of all time, and in it the laws of mechanics as seen to work on earth are related to the motion of heavenly bodies. The work of Newton put the physical sciences well in advance of all other branches, and as far as biology was concerned (Fig. 1.1) such vast unifying concepts were not to be produced until the mid-nineteenth century.

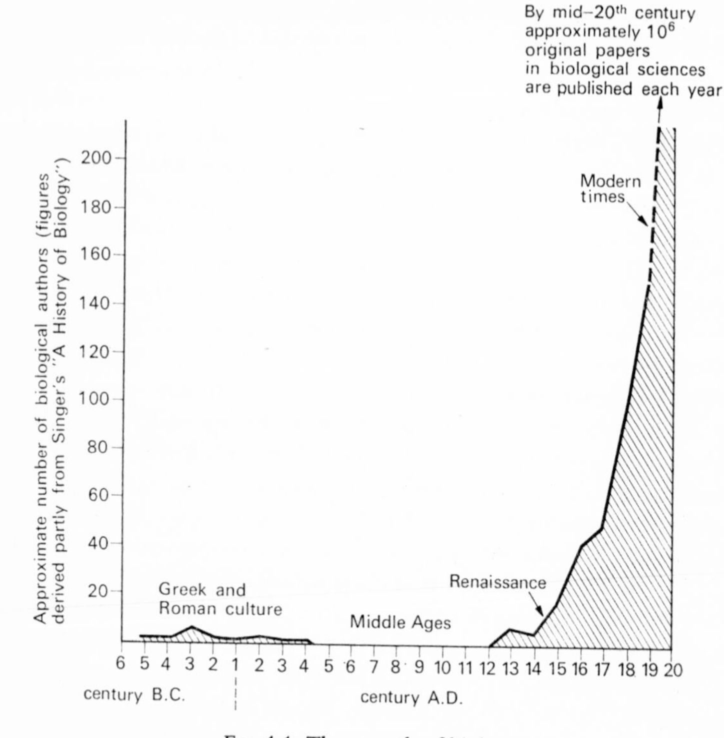

FIG. 1.1. The growth of biology.

The Methods of Modern Science

From the foregoing account the reader should have gained some idea as to the origins of the experimental and mechanistic analysis used in modern science.

In his book *The Art of Scientific Investigation*, Professor Beveridge discussed variously the roles of observation, reason, intuition and imagination, chance, hypothesis formation, and experimentation as all contributing to scientific discovery. Some of these are, of course, qualities of character or temperament.

To illustrate exactly how a scientific method can work we may take an example from biology, the investigation of the disease beriberi. (It should, however, be understood that the "methods" of cosmologists, modern chemists, molecular biologists, or theoretical physicists may be substantially different and much more involved than the instance quoted. Indeed, there is sometimes no definable method in discovery—particularly in introduction of totally new ideas.)

We start, then, with the noticing of a group of symptoms which seem characteristic for a particular disease which we may call beriberi. Because we know that diseases are often caused by germs and because we find outbreaks of beriberi in specific communities we suggest that this disease is the product of an invading microbe. This, then, is our hypothesis, and to test it we must discover the microbe present in those who suffer from the disease. When we find it we must culture it in pure form and on introducing it into a healthy animal we must find an attack of the disease develops. If indeed all this is achieved, we can conclude that our scientific hypothesis was correct and we have discovered the scientific explanation for the disease beriberi.

Thus observation along a particular line or into a particular phenomenon is followed by the setting up of hypotheses which may then be tested experimentally. If our hypothesis seems to fit the experimental results, then it can be used to predict what will happen in a given set of circumstances (e.g. that injection of a specific microbe will cause a certain disease). If this also works, the hypothesis is taken as correct and gains the status of a theory.

Now it so happens that the above procedure was indeed carried out on the disease beriberi in the last century, and in fact no infectious agent could be found in the bodies of sufferers. This might have been due to an agent being of ultra microscopic size, as happened in the case of yellow fever, but it could also have meant that the original hypothesis was incorrect. Going on this latter assumption, two workers, who had observed that beriberi is found where much polished rice was eaten, suggested that the disease was nutritional and was due to the lack of some vital factor from the diet (see p. 75). In the experiments it was found possible to produce the very symptoms of beriberi in chickens by feeding them on polished rice. It was further found that the sick birds were cured if the germ of the rice, previously denied them, was added to their diet. The general causes of beriberi were thus established, although it was a number of years later that the actual vitamin B which was lacking was found and only very recently that its actual role in metabolism was determined.

Important in all forms of experimentation, especially biological, are the setting up of controls. One must be quite sure that one is really investigating the particular factor that one thinks. Blank sets of experiments with everything present, except the factor in question, must be run. If you can produce beriberi in ten chickens fed exclusively on polished rice, ten similar chickens kept under the same conditions but fed on a normal diet must be kept and observed at the same time. If this was not done there could be a number of possible explanations for the occurrence of the disease.

The Main Themes of the Biological Sciences

As we have seen, the earliest applications of science were what we now call physics and astronomy. Chemistry was for many centuries encumbered with the mystic arts of alchemy, and much time was spent in searching for elixirs and the fabled philosophers' stone which would transmute base metals into gold. It was only in the seventeenth century that the quantitative methods necessary for real understanding began to be applied to the study of chemistry.

While it is true that much descriptive biology does not depend on assistance of the other sciences, the frontier researches of modern biology are almost entirely a matter of applying physical and chemical techniques to biological material.

Ultimately the phenomenon of life rests on the chemical structure and physical and chemical functioning of protoplasm.

One of the major themes of biology has been the description and classification of plants and animals into an ordered scheme—a scheme that was seen more latterly to exist by virtue of the process of organic evolution. This concept is the great unifying one of the subject, and most of the facts of biology can be related to evolution in some way or another.

Closely coupled to evolution, indeed a part of the process, is the study of heredity which includes the way in which variations arise and how they are passed on from one generation to another. The recent deciphering of the code of the gene is perhaps one of the most exciting of all scientific discoveries.

A further large part of biological work has been related to man and his anatomy and physiology. Much of the knowledge gained has been used in medicine, and this latter subject, together with the study of disease and its causative agents, are also central topics of biology.

Finally, there is the study of the plants and animals upon which we depend for our survival and the application of science to agriculture. The difference in yield obtained by the modern farmer from that obtained by his medieval predecessor underlies the whole of our economic prosperity.

Each of these topics, classification, evolution, heredity, medicine, and agriculture has certain key events and stages of advance in its history and certain individuals who have made great contributions. It is these events and their originators that will concern us in the following chapters.

Questions on Chapter 1

1. Who were the schoolmen and what part did they play in the revival of knowledge in the late Middle Ages?
2. Describe some of the origins and the importance of the experimental method in science.

CHAPTER 2

How Plants and Animals Are Classified

The Need for Classification

Half-way through the twentieth century some $1\frac{1}{2}$ million animal and 250,000 flowering plant species had been named and described, and the numbers increase annually. Even before there was the remotest idea that the various species could be grouped together in a way that indicates their evolutionary affinities, it was necessary to make some sort of ordered description of their multitudes. Man is a tidy, orderly creature himself, and likes to arrange things in patterns and levels. Indeed, making recognizable concepts out of disorder is one of the ways in which every human mind learns to make sense of the world around it. The great systematist Linnaeus not only arranged in order plant and animal species, but minerals, diseases, and even his friends.

Many animals and plants had properties that were useful, and so it was important to give them accurate descriptions, and eventually there was devised the efficient method of using the latinized scientific name for the same species wherever it occurred.

The idea of a species as a natural grouping of organisms was inherent in the Jewish stories of creation whereby species had issued immutable from the hand of God. The Father of Biology, Aristotle, discussed the principles of a physiological grouping of types in his *Historia Animalium*, but did not attempt to draw up a detailed system. Centuries later John Ray was to define species as groups of animals or plants which reproduce their peculiarities from generation to generation. Today the definition of a species as a naturally interbreeding group of organisms which can produce fertile offspring is only partly satisfactory. In fact, and as we shall see, the modern concept of a species is of necessity very loose, as species change with time and many show degrees of hybridization with others.

There are various possible sorts of classification, but probably the best from a biological point of view is based on a consideration of all the features of the organism being taken in due perspective which leads to a "natural" classification. This implies that one should be able to establish the criteria for judging the "due perspective", itself no easy matter. If we take a simple illustration the difference between a natural and artificial classification can be made clear.

Bees, caterpillars, ants, and flies are classed as insects, swallows as birds; and bats as mammals. We do not class together bees, flies, swallows, and bats as flying animals; such a procedure based only on one analogous function, that of flight, would be artificial and biologically misleading. Many early classifications were of this type, plants in particular being classified according to their uses. Of course, artificial arrangements have their uses and we shall see (p. 22) that natural systems may have severe limitations. Even in biological keys used for the identification of species, a considerable degree of artificiality is used. As we depart from biological to utilitarian classifications, the groupings may not be natural at all; thus shellfish to the fishmonger includes a whole range of quite different animal types.

Other mistakes in early biological classifications were made by taking a single feature, in fact one produced by parallel adaptations, as being of undue importance. Such was the case of the Radiata (radially symmetrical animals) all lumped together by the early systematists but subsequently recognized as belonging to many completely different biological groups such as Coelenterates (sea anemones), Echinoderms (sea urchins and starpoles) and Tunicates (sea squirts). Similarly, the many types of worms which were called collectively Vermes.

The Work of Aristotle and Theophrastus

Aristotle, the so-called Father of Biology, was born in 384 B.C. when Greek civilization and learning was at its zenith. Not only was he a remarkably astute biologist but also a philosopher, studying under Plato and later founding a school, the Lycaeum, of his own.

As far as biology was concerned he had the opportunity of investigating marine animals on the shores of the Aegean Sea, and described the anatomy and functioning of many species, writing and teaching until his death in 322 B.C.

In his writings he attempted to group together certain natural objects and arrange a ladder of nature. He distinguished minerals with no "souls"* from vegetables which he thought had souls which only allowed nutrition, growth, and reproduction, while animals had higher souls which were capable of movement and response. At the top of his tree came man—literally, for woman was placed second—whose soul allowed him to reason. (The fragments of this idea are still seen in the children's game of animal, vegetable, and mineral).

Aristotle thus arranged nature with man at its summit and inanimate matter at its base, and some modern authors—both scientific (Hadzi, 1953) and philosophical (du Chardin, 1958)—are still doing the same.

The major consideration in his scale was given to reproductive methods and to blood colour as well as the degrees of soul described above. At the top of the animal kingdom he placed the animals with red blood and young born alive and included in descending order mammals, the cetacea (marine mammals such as whales), reptiles, birds, amphibia, and fishes. Below these he placed animals with colourless blood but "perfect" eggs— the cephalopods (squids and octopi) followed by crustaceans, then insects, slugs, and bivalves in a single group. Oddly enough he thought worms were the young forms of insects and must have been confused by the vemiform shape of caterpillars. Lower down came the animals with imperfect reproduction, "spontaneous generation", budding, etc., which included the jellyfish, corals, sea squirts, and sponges. Below these were the plants about which he wrote much less, mainly grouping them according to habit.

We may laugh at the mistakes he made, but in fact Aristotle knew only some 500 species of animals, and his work is a valiant and in

---

*"Soul" here is equivalent to "psyche", and means a state of being.

many ways skilful attempt to organize his knowledge into a logical pattern. Among systematists he is particularly admired for his internal arrangements of the molluscs and fishes.

A pupil of Aristotle was Theophrastus, and he did much for plants that his master had done for animals including an attempt at their classification (Fig. 2.1). Theophrastus distinguished flowering from non-flowering plants and even between those that had two embryo leaves in their seeds (dicotyledons) and those that had one (monocotyledons). He also recognized the difference between fused- and separate-petalled flowers and the relationship of the ovary to the rest of the flower—all important points of modern taxonomy. In his works are included many details about the medicinal uses of plants.

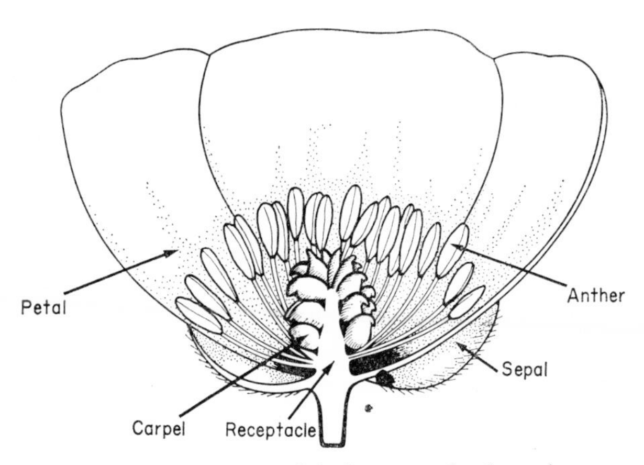

FIG. 2.1. A half flower of the buttercup family to show the floral parts used in classification.

Predecessors of Linnaeus

In the Roman culture that followed the Greeks, little work was done on zoology apart from a collection of legend and nonsense by Pliny which unfortunately also became incorporated along with the really great Greek scientific literature into established authority.

Botany fared better as the use of herbs by doctors in the Roman army demanded reasonably exact methods of their identification, and in the first century A.D. a physician, Dioscorides, produced a herbal for the specific purpose of plant recognition. In this certain natural groups such as the Deadnettle and Carrot families were distinguished.

During the Dark Ages that succeeded the Roman civilization, little new work was done and the biological writings of the classical authors were slavishly transcribed. So ridiculous was this situation that for drawings of medicinal plants in northern Europe scribes copied pictures of species that existed only in the regions of the south rather than drawing appropriate similar species that grew on their own doorstep. It was during these centuries that "beastaries", which included many strange and mythical creatures such as griffons, as well as real ones, were put together. In general the properties of the beasts were related to various human or divine attributes either as a warning or example.

After 1500 a succession of men worked at biological classification and the description of animals and plants. Notable among the botanists are Cesalpini (1519–1603), Brunfels (1489–1534), Boch (1498–1554), and Fuchs (1501–66), all of whom published new descriptions of the plants known to them and which had not been described by the ancients. For animals Gesner (1516–65) produced a copious *Historia Animalium* and others wrote on birds, insects, and some of the new material, still partly legend, from the New World. One of the most comprehensive of these was a work on the flora and fauna of Mexico published by the Academy of the Lynx (see p. 7).

There seems throughout the seventeenth century to have been a growing interest in all aspects of nature and natural science, and many scientific societies — including the Royal Society — were formed at this time. By 1620 a book by the Swiss Bauhin describing some 6000 plant species was produced in which some distinction between genus and species is made, and botanists were reaching out to the true nature of plant organs as a means of determining their correct arrangement. Conspicuous forerunner of Linnaeus was John

Ray (1628–1705), the botanist who produced the first flora of his local Cambridgeshire, then of all the plants of Great Britain. In this latter work Ray organizes some 18,000 plants by a method based on a better understanding of their sexuality than existed previously; and by taking all the features he could into consideration he achieved a more natural classification than his predecessors. Besides his botanical publications Ray also worked with Willughby on animal classifications, and among the mammals they were able to distinguish single-hoofed (now called perissodactyls) from the double-hoofed (artiodactyls) and these from groups possessing claws. Within the latter they recognized the carnivores, rodents, elephants, and monkeys.

The way was now paved for the great binomial classification of Linnaeus which in many respects is the same as that used today.

Linnaean and Other Pre-Darwinian Classifications

As already mentioned, the Swede Carl Linnaeus (1707–78) had a passion for classification and a great deal of talent in distinguishing categories and inventing systems. Despite the fact that he believed species to be fixed or capable only of very limited hybridization, he succeeded in making in many cases a natural classification at a generic level and below, and like John Ray he took many characters of an organism into consideration when fixing its systematic position. Ironically enough the fact that natural classifications can be constructed at all is now taken as evidence that evolution has taken place, but it is quite possible to construct such classifications without knowledge of evolution and this is what Linnaeus did.

The binomial system of nomenclature used by Linnaeus was not, in fact, invented by him but was certainly popularized and made acceptable by his adoption of it.* By this system every animal and plant was given two names, the first being a generic name which might apply to several closely related organisms, and the second a descriptive specific name. The combination of the two names would

---

*The Swiss Bauhin (1560–1624) probably invented the binomial method but he did not use it consistently, and its effective introduction was due to Linnaeus.

be unique, thus *Homo sapiens* for man, *Felis leo* the lion, *Felis tigris* the tiger, *Bellis perennis* the daisy. The first name has a capital letter and the second a small one, and nowadays it is customary to include the abbreviated name of the classifier. Man thus becomes *Homo sapiens* Linn.

Linnaeus arranged his species into related groups which he called orders and these into larger classes. The latter he placed in kingdoms which were major groupings. Since the publication of his great work on classification (the *Systema Natura*, first edition in 1735) other categories have been added to the original Linnaean system such as family for sub-divisions of an order, and phylum for a collection of related classes.

The full classification of man is as follows on the modern system.

Kingdom	Animal
Sub-kingdom	Metazoa (many celled)
Phylum	Chordata (hollow dorsal nervous system, notochord, gill slits or equivalent structures)
Sub-phylum	Craniata (skull and backbone)
Class	Mammalia (warm blood, fur or hair, mammary glands)
Sub-class	Eutheria (placenta present)
Order	Primates (climbing, rather primitive, opposed thumb)
Sub-order	Anthropoidea (great apes, monkeys, and man)
Family	Hominidae (types of man—include several fossil forms)
Genus	*Homo* (man)
Species	*sapiens* (the wise one)

For his arrangement of plants Linnaeus paid attention to the number of stamens and stigmas possessed in the flower, and his classification is known as the "sexual system". The *Species Plantarum* (1753) contained 6000 species in 1000 genera, and although this major stress on a single feature leads to a rather artificial system, it at least allowed flowering plants to be readily identified. A further excellent feature of Linnaeus's work was his invention of a concise terminology for the organs of plants and animals which greatly facilitated description.

The plant classification of Linnaeus was as seen in Fig 2.2.

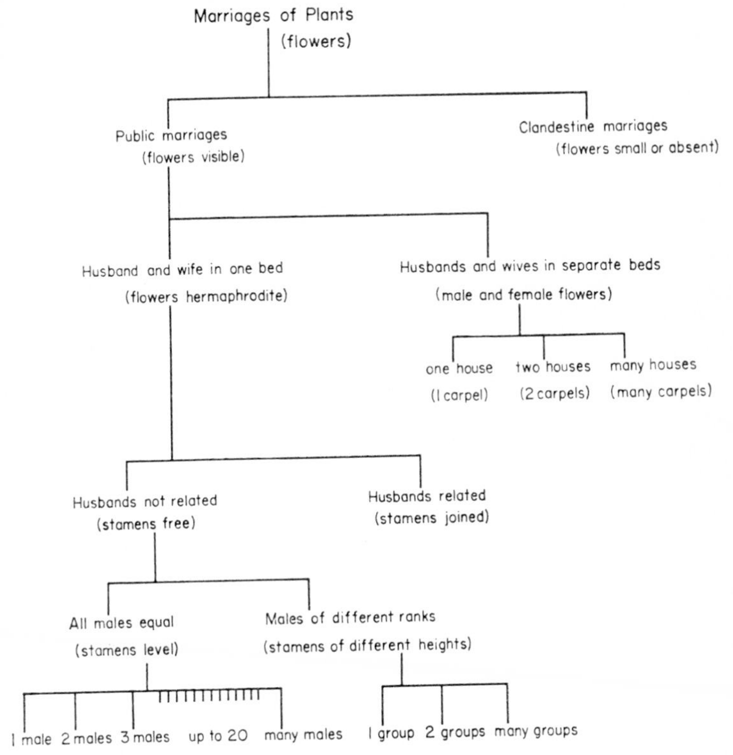

FIG. 2.2. The way in which Linnaeus arranged flowering plants. This should be compared with Bentham and Hooker's scheme (Fig. 2.3).

As far as animals were concerned, the various classes of vertebrates were distinguished correctly as the Mammals, Birds, Reptiles, Amphibians, and Fishes using the anatomy of the heart among other features. Within the group he was less successful—of the eight orders of mammals he described only three are used as natural groupings today.

With the invertebrates he lumped great classes together as Insects or Vermes, and his system has been much revised.

Despite its many mistakes and shortcomings the biological classification of Linnaeus was far better and more comprehensive than any previous one, and his concise method of description was a model for later biologists to follow.

Following Linnaeus's admittedly often artificial schemes of biological order, his successors in both botany and zoology strove more and more towards a natural classification.

In botany de Candolle published a comprehensive flowering plant classification of 60,000 species (1824) based on the idea that primitive flower families have many free parts and fusion is a sign of advance. He thus placed the Buttercup family at the base of his scheme and, as it happened that some of his ideas coincided with the majority of evolutionists, his scheme was to influence post-Darwinian classification. The final pre-Darwinian classification of plants that we shall consider was made by Bentham and Hooker between 1862–83, and although these authors published after the *Origin of Species* their work had been done without a conscious attempt to

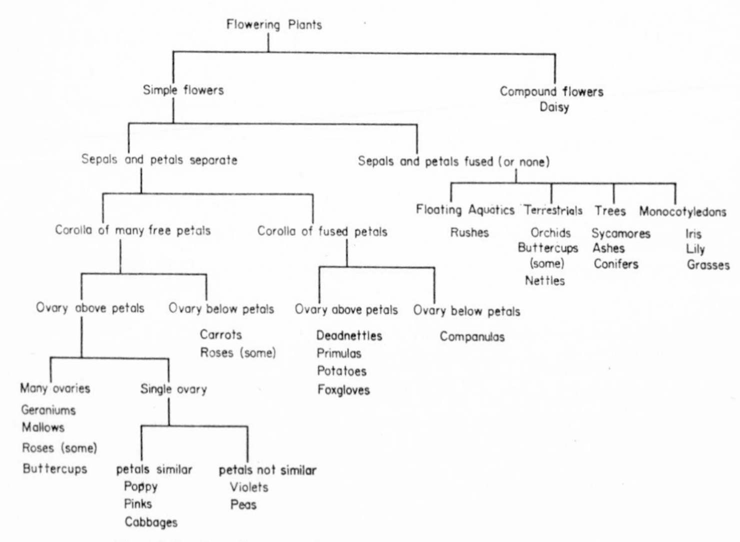

FIG. 2.3. Bentham and Hooker's key for identification of flowering plants (from 1904 edition).

produce an evolutionary system. Their classification of flowering plants was as in Fig. 2.3.

A rather modified version of the above scheme is still used in many British floras of the present day.

For animal classification the works of Lamarck published in 1809 and Cuvier in 1817 are important landmarks. The former used characters of circulation and nervous system and wrote a very large classificatory and descriptive catalogue of animals in which he recognized the backboned animals, the molluscs, the articulated animals, and the radial animals. He did not appreciate the status of the Amphibia, and in the lower animals he included the barnacles (crustacea) with the molluscs and the annelid worms with the arthropods.

Lamarck introduced the term invertebrate to include all the lower animals and was successful in separating these into many of the various phyla that we recognize today. These groups of Lamarck's were the Molluscs (snails), Cirripeds (barnacles,) Annelid worms, Crustaceans (crabs, etc.), Arachnids (spiders), Insects, Worms, Radial animals (starfishes, etc.), Corals, Infusoria (single-celled animals). He arranged these in order and believed that evolution had taken place from one stage to another. Cuvier believed entirely in special creation and the fixity of species. Curiously enough it was Cuvier who established that fossils are the remains of living organisms, but his study of geology led him to interpret the collections of fossils he found as being the results of recurrent deluges or other natural catastrophes.

Modern Systems of Classification

As we shall see in the next chapter, the old idea of the fixity of species was to be shattered in the mid-nineteenth century by Charles Darwin's demonstration of their origin by natural selection operating over vast periods of time. The concept of organic evolution was to influence all biological thinking and not least that concerned with systems of classification.

The most obvious implications of a belief in evolution was an explanation of how the diversity of living things had come about and to suggest that the degree of resemblance indicated the degree of evolutionary relationship between two species. It was not surprising that the aim of classificatory or taxonomic work became the accurate expressing of such evolutionary relationships, and phylogenetic or evolutionary classifications were made.

A hundred years after the demonstration of evolution we find ourselves not quite so certain about the absolute merits of phylogenetic classifications. In the first place an unquestioning adoption of this aim for taxonomy obscured the fact that even if living things had been specifically created they would still have needed to be classified and, indeed, had been repeatedly classified long before evolution was thought about. Soon after Darwin published his *Origin of Species*, T. H. Huxley was to make the point that there can be other and equally worthy aims and purposes to classification besides the demonstration of evolutionary affinities.

At the present time there appear to be two distinct schools of thought about biological classifications, and in considering the nature of modern work in this field it is as well to be aware of this fact. The first, and in a sense, more conservative school, holds the view that the aim of classification is to express evolutionary affinities more and more accurately as new facts emerge and to believe that the most perfect scheme is the one that does this most completely. The other school of thought argues that there are many valid reasons for classification, and the expression of phylogenetic relationships is only one of them. This school, self-styled the "doubters", would advocate broad general purpose classificatory systems which serve the greatest number of uses.

Although the former ideal has an obvious and attractive motive and possibly represents the goal of most taxonomists, there is, in fact, much to be said for the latter system. In the first place a classification should be of use to somebody—this is a primary object of classifying anything at all. Thus to a field ecologist the separation of certain plants into calcicoles (chalk-loving) and calcifuges (chalk-hating) may be much more useful than some phylogenetic scheme based on the families to which they belong. Again, a marine biologist may find it more useful to use an arbitary quality such as colour or bristle number or even type of burrow in constructing an identification key than some more deeply meaningful evolutionary quality.

Another and extremely important point is the difficulty that is involved in actually constructing a complete evolutionary classification. Using all the up-to-date knowledge of genetics it is hard enough to determine the relationship between two similar species, and when we go further back to genera, families, orders, and classes, the difficulties are almost insurmountable.

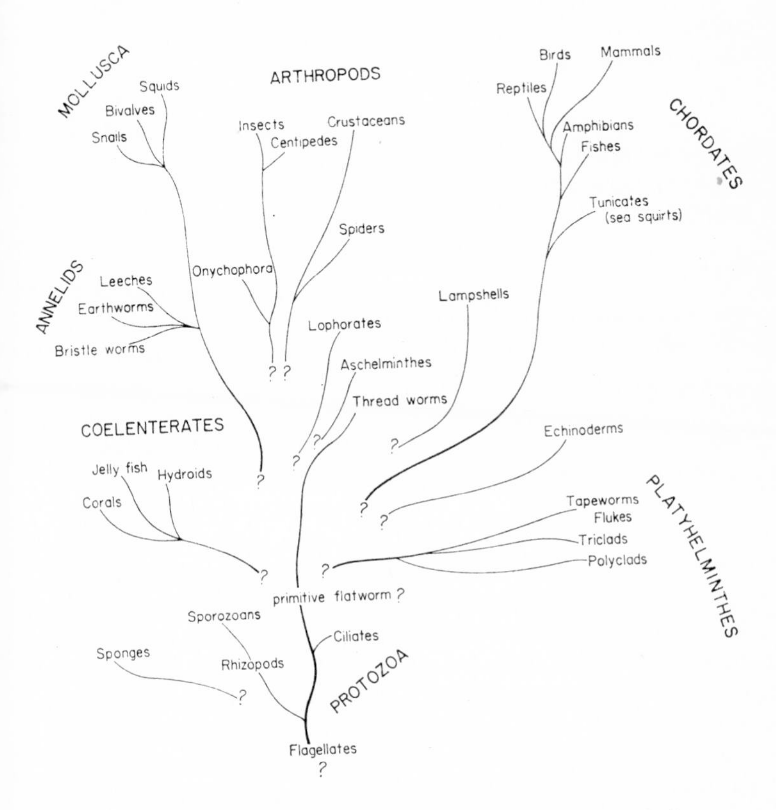

FIG. 2.4. A possible phylogeny of animals (after Hanson, 1961).

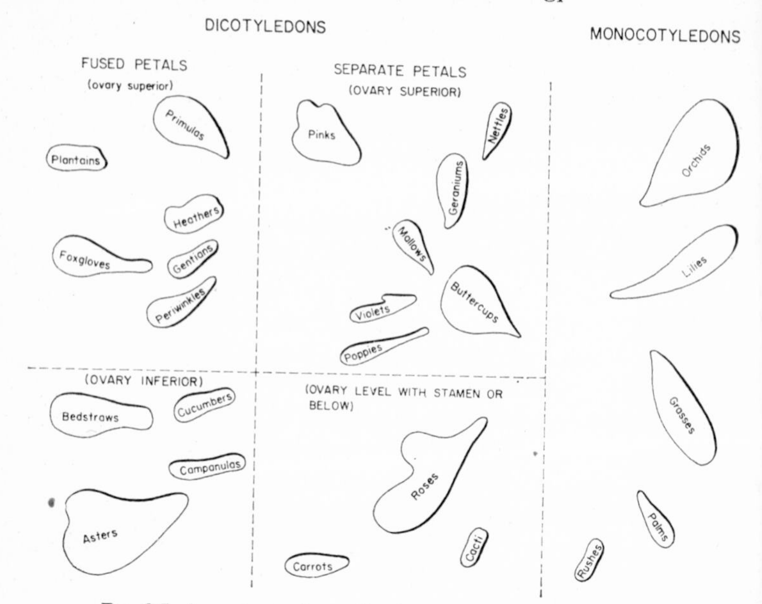

FIG. 2.5. A modern scheme for showing the affinities of some flowering plant families. There are no connecting lines drawn, but proximity is an indication of probable evolutionary relationship.

A recent outline classification of the major phyla of the animal kingdom is as in Fig 2.4, and the affinities between one phylum and another is shown. In his *Implications of Evolution*, published at about the same time (1960), G. A. Kerkut produces reasonable alternatives for the origins of most of the main phyla. It all happened a long time ago; there are no fossil intermediates left and we just do not know exactly how the pieces fit. Whatever the ideals of our classification there seems no particular reason why we should know.

As far as plants are concerned the position is equally confused, and the production of a phylogenetic classification has proved immensely difficult if not impossible. Again, we can see an up-to-date representation of the relationships between various flowering plant families (Fig 2.5) where the shapes represent sections of a series of evolving lines. The shapes of the lower branches of this "tree" cannot be drawn in.

Despite all this and the difficulties, there is still much work done on taxonomy. Certain aspects of this work concern themselves with the units of population on which the processes of variation and selection can be observed. Such studies were called by Julian Huxley, grandson of the T. H. Huxley mentioned above, the "New Systematics" (1940) and prominent in this field has been the work of Dobzhansky (1937), who studied the genetic divergence of populations, Simpson (1944), who worked on evolutionary rates, and R. Fisher, who worked out the mathematics of gene selection. From this sort of approach we have learned much about the working of the evolutionary process and the nature of species formation.

The old pre-Darwinian idea of classification was to arrange the diversity of nature in order. Now the emphasis has shifted, and as we have come to a deeper understanding of how this diversity arose so our objectives have become more limited but at least more capable of realization.

Questions on Chapter 2

1. Aristotle arranged natural objects such as plants, animals, and rocks into groups according to the amount or degree of "soul" they possessed. What groups did he distinguish and what did he mean by "soul" in this context?
2. Why do we find it necessary to classify plants and animals? Outline the scheme of classification used by Linnaeus.
3. Since the advent of Darwin the scientific notion of species has become much more complex than before when it was assumed that they were created as such and could not change. Discuss.

CHAPTER 3

Evolution

Early Thinking on Evolution

The central theme of biology is evolution. This implies that living organisms have emerged from the sub-living organic level and by a process of selection and change diverged into the great variety of plants and animals that have flourished in the past and are alive today.

The principle of evolution applies just as much to man as it does to all living things, and it is an important part of our understanding of our place in nature to know our origins and the forces that have brought about the present state of our minds and bodies.

The first major difference between our modern theories of evolution and earlier notions is the vast quantity of time involved. The palaeontological or fossil record shows us innumerable forms of living being, most now extinct, slowly changing from level to level into our earth's present flora and fauna. There is a sense here in which man, a late-comer in time and just one evolving line among tens of thousands, gets a new perspective of his position in nature.

A second important idea of modern theory is the randomness of change. Mutations, the spontaneous gene changes upon which natural selection operates, appear to arise in no clear pattern, and, indeed, are mostly harmful rather than beneficial. An end or direction in evolution is not an idea favoured by biologists although they may recognize that levels of complexity and specialization exist. Thus certain methods of organization are more efficient and less open to destruction by hostile environmental forces than others, so that we may call creatures with these elaborate controls "higher" as compared with "lower" organisms.

We recognize that many-celled animals are more advanced than unicellular, that animals with intelligence are more advanced than those without, that those whose physiological mechanisms render them immune from externally applied changes are advanced beyond those that have no such mechanism. In this sense, then, man does represent some sort of evolutionary peak. Despite his comparatively unspecialized mammalian body, his highly specialized brain and behaviour set him apart from other creatures, and the activity of his mind acting in his social environment causes new evolutionary forces to come into play. Our species, emerging by the slow biological processes of mutation and selection upon which all other species depend for variation and change, is partly emancipated by a new upthrusting force. This latter is the continuity of mind from one generation to another, the collection of ideas and discoveries, as if they were gene mutations, into the human "social-plasm". We are no longer born into the same unchanging world of our ancestors, each new generation of man, "the thinking one", spends the greater part of his life in a cultural environment unknown to his predecessors.

The speed of human and social change and cultural evolution far exceeds that of biological change. A modern educated man and a Neolithic hunter may be father and son as far as their biological features are concerned. In the world in which they live and in which their thoughts move, they are separated by a void such as exists between two distinct species.

Despite the ladder of nature arranged by Aristotle (see p. 14), this was in no sense meant to indicate an evolutionary relationship between one rung and another. Anaxagoras (*c.* 499–428 B.C.) had suggested that matter as we know it is made from particles or seeds from infinite space put together by an external intelligence and at death resolved back into particulate form. This notion was refuted by Aristotle who was much more of a vitalist than evolutionist in outlook, and his idea of the soul or psyche (described in Chapter 1) was the basis of the vitalistic physiology that existed in the nineteenth century. Vitalists regarded living organization as in some way special and not subject to complete mechanical analysis.

Against any notions of organic evolution stood the firm bastion of the creation story handed down in Hebrew writings. In the authoritarian doctrines of Roman and medieval Christianity, species had been created individually at the beginning of the earth by God according to the accounts in Genesis.

Later philosophers toyed with the invention of ideal systems which employed the concepts of improvement and change. Linnaeus in 1735 had produced his great classification of nature which, despite its biologically sound framework, was made by him without any knowledge of the real affinity by descent that linked his species. In fact it is supposed that his own attempts to classify nature had eventually led to him believing in the genus as the created unit and the species as a more variable form resulting from interbreeding.

It is strange that a more or less correct evolutionary classification —now seen as one for the major supports for the fact of origin by descent—existed long before its significance was realized.

However that may be, the classification of plants and animals need not lead to an acceptance of evolution and, as we saw in Chapter 1, many steps were taken in the former while the latter was undiscovered. From a very wide knowledge the French biologist Buffon (1707–88) had come to realize that animate nature was only a part of that whole universe, some of whose laws and physical continuity had been described by the astronomers. Buffon did not favour the idea of fixed species but saw all things in a state of flux so that any classification was an arbitrary statement of a temporal situation. He also studied fossils as indicators of earth history and of biological chronology. This was a far cry from the idea that fossils were strange "sports of nature" or the remains from Noah's flood current in medieval literature. Like others before him, Buffon was much impressed by the homology of organs and realized clearly that vestigial parts, such as the splint toes of certain mammals, were degenerate by a process of modification from a former "perfect" state.

We shall find many threads lead to Charles Darwin—at least one comes from Buffon via Erasmus Darwin, the grandfather of Charles. Although he was rather more of a theorist than a practical biologist, Erasmus Darwin came to the conclusion, writing in 1795, that animal species are the result of change and divergence from common ancestors, the divergence being caused by the surroundings which they inhabit. He was particularly impressed by the divergence which man has caused in certain domestic animals such as dogs, horses, and pigeons by selective breeding for specific qualities.

Meanwhile in France a young contemporary of Erasmus Darwin called Lamarck was producing fine classifications of the invertebrates, amplifying and sorting out some of Linnaeus's "lump" groups such as the Vermes and Arthropods. The more he learnt about living things the clearer he became in his own mind that so-called species were transitory forms and that an "evolutionary process" had produced diversity. Thus Lamarck preceded Charles Darwin in the belief in the fact of change and the incorrectness of the fixity of species doctrine.

As Darwin was later to do so successfully, Lamarck tried to reason how such change had come about. His conclusions were that life by its own nature grew more complicated and that new needs evoked new organs. He also thought that use of an organ or faculty led to its development and disuse to its reduction (as indeed is true within the life of an individual) and that these changes were passed on to successive generations. The former ideas are not of a scientific nature and cannot be tested by scientific analysis, but the latter notion is known as "the inheritance of acquired characters", and it is for this that we remember Lamarck. Though inheritance of acquired characters was broadly refuted by Weismann (see p. 46), *neo-Lamarckism*, by which is meant the tendency of a species to produce the "correct" adaptation to environment—in some fashion that does not involve random mutation—still exists as a school of thought in present times. (There are also cases of inheritance of acquired characters such as behaviour in flatworms and characters of tree grafts, but these are exceptional and most investigations show that inheritance of acquired character does not normally occur.)

Darwin and Wallace on the Origin of Species

In the closing years of the eighteenth century the topic of evolution was much in the air. A series of important geological works by

Hutton (*Theory of the Earth*, 1795), William Smith (*Stratigraphical System of Organised Fossils*, 1817), and Charles Lyell (*Principles of Geology*, 1830) made available a solid corpus of knowledge about the sequence of geological and certain biological events in the earth's history over long eras of time.

Charles Darwin's contribution to biology was to be twofold. He was to accumulate evidence to show that evolution had actually occurred, but more important he was to present a means whereby it *could have occurred.* In describing his mechanism of evolution Darwin was influenced by an *Essay on Population* by Malthus published in 1798.

In this essay Malthus points out that for man prosperity and good conditions cause a rapid increase in the population—in fact a geometrical increase. (Geometrical increase is common to all biological systems and involves a doubling in each generation—thus 1, 2, 4, 8, 16, 32, etc.—it means that the rate of increase itself increases with time.) Food supplies, however, tend only to increase arithmetically, i.e. by a fixed amount (1, 2, 3, 4, 5, 6, 7, etc.). Quite obviously a factor increasing geometrically will soon overtake one that expands arithmetically, and applying this to population Malthus showed periods of increase would be followed by famine and other checks.

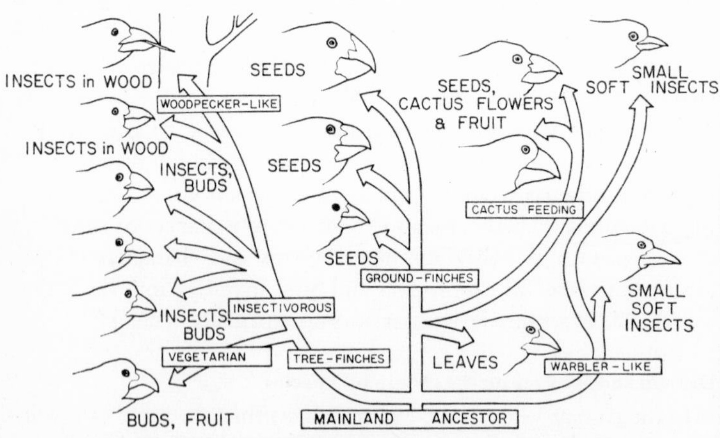

FIG. 3.1. The divergence of Galapagos finches as described by Darwin.

The young Charles Darwin had in 1831 accepted the post as naturalist aboard H.M.S. *Beagle*. During his trip around the world Darwin had collected fossil mammals from South America whose affinities with living forms were obvious to him. Here, perhaps, he grasped the truth of the change of species in time, and on his arrival at the Galapagos Islands, 500 miles out in the Pacific, he seems definitely to be collecting evidence to show such changes had occurred. On these islands Darwin noticed the changes in tortoises and finches (Fig. 3.1) and other species from one island to another and here, and throughout the remainder of his trip he amassed a great body of facts together which he was later to organize into reasoned support for his theory of evolution.

Two years after his return home Darwin happened to read the essay by Malthus on population mentioned above, and it immediately occurred to him that in any struggle for survival, favourable variations would tend to be selected. This gave him a theory to work on to explain the divergence of species and by 1858, 20 years later, he had written up his copious data to support his theory of evolution by natural selection.

It was in this same year that Alfred Russell Wallace sent Darwin a copy of his paper 'On the tendency of varieties to depart indefinitely from the original type' which paralleled Darwin's own views and led to a joint statement to the Linnaean Society on evolution. In the following year the *Origin of Species* was published in which evidence for Darwin's theory was given in full.

His major ideas can be summarized thus:

1. Both individual organisms and their offspring show variations.
2. Those organisms that have favourable variations will survive to reproduce their kind, while those with unfavourable variations will perish.
3. The perpetuation of such variations leads to a gradual divergence from the original stock. In this way species and eventually genera and orders are formed.

Of course the whole theory ultimately depended on the source of the variations about which Darwin knew nothing (see p. 31), but even as it stood the theory did provide a workable means whereby divergence could have occurred by a *natural* process.

To support the fact that evolution had taken place at all, Darwin produced geological evidence, evidence from the homology of structures such as the limbs of vertebrates, evidence from the geographical distribution of animals and from the results of human breeding of animals. To support his theory as to the means whereby it had occurred, he showed the large numbers of offspring produced by many animals and demonstrated the competition that existed for survival in nature. Effective work on variation and heredity was not to be done for a further half-century.

Thirteen years after the *Origin of Species*, Darwin brought out his *Descent of Man* in which the general thesis of evolution was applied as far as was then possible to man himself, and his affinities with the higher primates was clearly revealed.

The acceptance of Darwin's theory has only been brought about after years of altercation and by the ever-accumulating mass of biological fact that supports it. In actual fact no one has ever seen the origin of any species (except a few polyploid plants, e.g. *Primula kewensis*) because of the time involved in the selection process, but the fact of evolution and its mechanism is generally accepted by biologists as taking place along the lines laid down by Darwin and the later geneticists (Fig. 3.2).

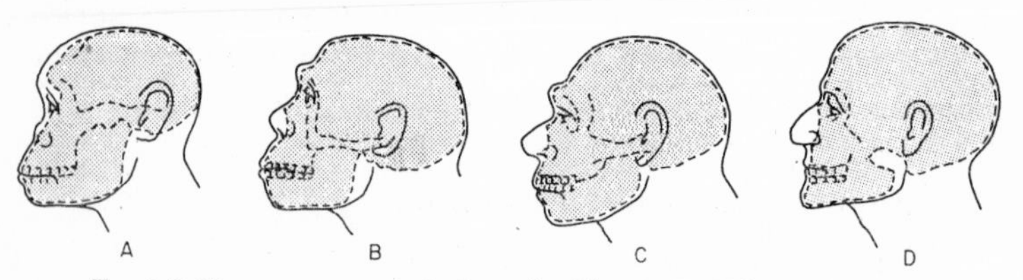

FIG. 3.2. Human ancestry. A, Australopithecus. B, Pithecanthropus. C, Neanderthal man. D, Modern man.

We are not here primarily concerned with the philosophical or other implications of knowledge, but it must be obvious that the Darwinian theory represented a complete new stage in human knowledge. The old ideas of the creation of fixed species were gone for ever, the vast age of the earth was revealed, and man given a new perspective in his place among living organisms. The repercussions and implications of evolution are still felt today in almost all fields of knowledge, and the world can never quite go back to the enchanted garden of Eden and the 7-day creation which Darwin clearly showed was mythological rather than actual.

Evolution since Darwin

The century of researches that has passed since the publication of the *Origin of Species* has done much to confirm and expand Darwin's idea of evolution by natural selection. Almost every field of biological investigation has yielded material that either relates to the fact of evolution having occurred or else to the mechanism by which it occurred.

The Fact of Evolution

Compared with our present-day accurate radioactive dating methods and the richness of the fossils known to palaeontologists, Darwin and his contemporaries had the crudest methods of dating and very limited material. Important from the point of evolution are the fossil series whereby an organism can be seen to change gradually over a long period of time to give rise to new, more highly adapted species. Such series have been described for the horse (Simpson) in its change from a small four-toed animal in the Eocene to its present large single-toed form, and it has even been estimated that this change involved some 15,000,000 individuals. It is also well seen in the ammonites of the mesozoic, coiled swimming molluscs whose shell patterns grew more and more complex with time, and in the echinoderms (sea urchins) of the same period. These two latter organisms are in fact used as zone fossils of the rocks laid down at this time, allowing comparison and identification of strata. Fossil series also can be traced in the trilobites of the Palaeozioc, early arthropod forms, whose anatomy showed gradual changes from their appearance in the Cambrian to their final extinction in the Carboniferous period.

A study of these series gives us a clear picture of the sort of jumps that are made by an evolving line and the rate at which change may take place (Fig. 3.3).

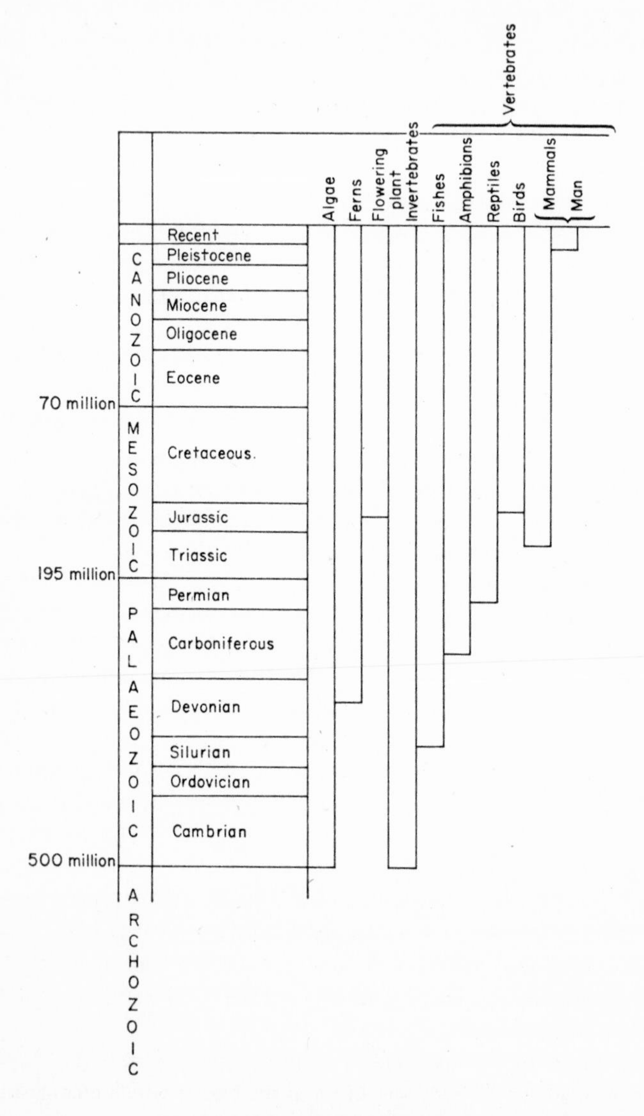

FIG. 3.3. The geological time scale.

The findings of "living" fossils and link forms of animals and plants have helped us fill in the evolutionary gap between major classes and phyla. Some of the more interesting of these are the Coelocanth (fish → amphibia), Seymouria (reptile → amphibia), Archeopteryx (reptile and bird), Ictidosaur (reptile → mammal), Peripatus (worm → insect), Neopilinia (worm → mollusc). It seems evolution may take place most rapidly in transition forms which will also tend to be less successful, hence less numerous than their descendants, so they are seldom found. The close fit between predicted transitional types and the actual ones discovered has been a further substantial vindication of the evolutionists.

In the field of comparative anatomy and physiology much work has been done which lends confirmation to the idea of divergence of species. It is found that within a single group of plants or animals such things as flower or limb structure show homologies and adaptive differences most easily explained by the assumption that evolution has taken place. Thus the pineal eye of vertebrates at first functioning as a photo-receptor (light-sensitive organ) still retains an eye-like structure even when recessed below the surface of the skull and functioning as an endocrine or hormone organ. There is much evidence that the biochemical substances used in certain metabolic activities show signs of evolutionary affinities. The group of invertebrate animals thought to be closest to the vertebrates have similar organic excretory products, and within the mammals confirmatory work on the relationship of the various orders has been provided by precipitin tests between blood sera (Nuthall, 1904). Recent biochemical work in the USSR and America has also furnished a clue as to the way in which life originated from inorganic matter—the very first but greatest step in the evolutionary process.

Only a few years after the publication of the *Origin of Species*, Haeckel (1866) produced a theory of embryological resemblance which purported to show that the embryonic forms of higher animals resemble the adult types through which they evolved. This was much modified subsequently by Garstang and other workers, and today it is thought that embryo forms may resemble previous stages of embryo rather than adult (Fig. 3.4). An interesting view is

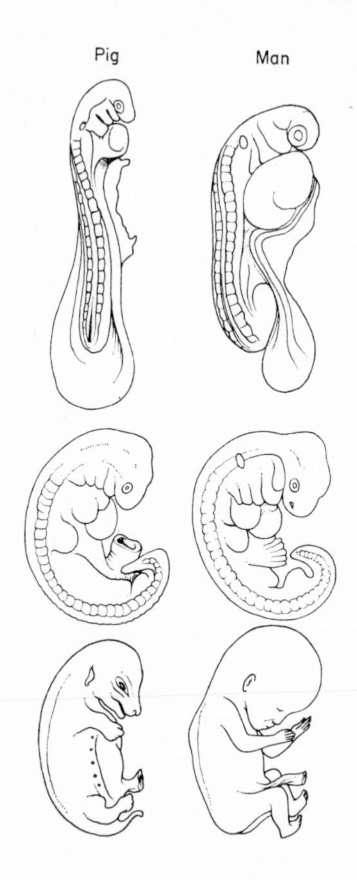

FIG. 3.4. Embryological affinities are well seen in mammals.

that certain major evolutionary jumps have been made by modification of a sexually mature embryo which has superseded a more static adult form. Such paedogenesis, as it is called, may indeed have been the origin of our own class of vertebrates from the sea squirts.

The geographical distribution of animals studied by both Darwin and Wallace has been extended and applied also to plants. It appears evident that South America was cut off from North America in the

Pliocene and has basically a very primitive mammalian fauna. Similarly, Australasia, with its living monotremes and marsupials, represents a still earlier isolation from the main body of mammal evolution.

The study of the flora and fauna of many oceanic islands has also confirmed the idea based on Darwin's survey on the Galapagos Island that evolution operates most rapidly under such conditions of isolation.

*Natural Selection**

It is always important to separate the idea that evolution has occurred from the mechanism of its occurrence. Although much evidence has accrued in the last century to support the former idea, the major advance since Darwin's time has been in regard to our understanding of the mechanism of natural selection.

Let us take some of the individual points of Darwin's theory. First he said that organisms increase more than sufficiently to maintain their numbers—an observation on the explosive nature of biological reproduction that has been confirmed again and again. In Australia the prickly pear cactus and the rabbit, introduced in ignorance, reproduced at such a rate that they filled up great areas of that continent with their millions of offspring. Similarly, in New Zealand brambles and cabbage white butterflies showed the effects of uncontrolled reproduction. In the Middle East and Africa great clouds of locusts in flights measuring some 14 square miles have been found to be produced from a few thousand individuals within a period of weeks. All these powerful explosions of particular species have been due to favourable factors and lack of competition.

For his second statement Darwin suggested that competition between offspring must occur, and many laboratory tests on animal populations have been made. In the 1930's Gause worked on populations of single celled animals and of fruit flies and observed the factors which terminated and controlled their geometrical increase within a limited environment. The obvious fact that competition

---

*This section should be read after the following chapter on heredity.

must take place between a large number of individuals that will eventually be reduced to a few was placed on a mathematical basis, and the intensity of competitive and selective forces in relation to the size of the population calculated.

The next point Darwin made was that the offspring of an individual vary. In fact he understood very little about the nature of variation, and it was not until the discovery of mutations by De Vries and subsequent investigations on their frequency and nature by Müller that true understanding was possible (see p. 53). Variations were seen to be due either to the reshuffling of genetic material that takes place in sexual reproduction or else to the actual chemical change of this material by mutation. The latter changes caused by radiation and other environmental causes were found to be mostly harmful but a small number to be beneficial, and these were the key factors of evolutionary advance. For most lower animals the mutation rate of a single gene may be 1 in 500,000, this rate varying in both higher and lower animals. Such a rate seems itself to have been selected giving sufficient rate of change for adaptation but not enough to cause constant break up of favourable genetic combinations.

The discovery, originally made by Mendel, that heredity units existed that did not blend, was a vital clue in understanding the mechanism of variation and its transmission to future generations. Had character variations blended then variation itself would have been lost and evolutionary divergence could not have occurred.

The point that Darwin understood little about the nature of inheritance has been made in the third chapter. Since the rediscovery of Mendel's laws of inheritance it has been known how such character inheritance takes place. For the determination of most characters many genes operate together, and in the 1930's Fisher showed that new mutations would tend to be inherited in such a way as to show their nature, i.e. to become dominant if they were beneficial and not to show it, i.e. to become recessive if they were harmful.

As to the operation of natural selection on variations, the critical point of the Darwinian evolutionary mechanism, the amount of practical work that it has been possible to do is still surprisingly small though many theoretical models were made (Dobzhansky).

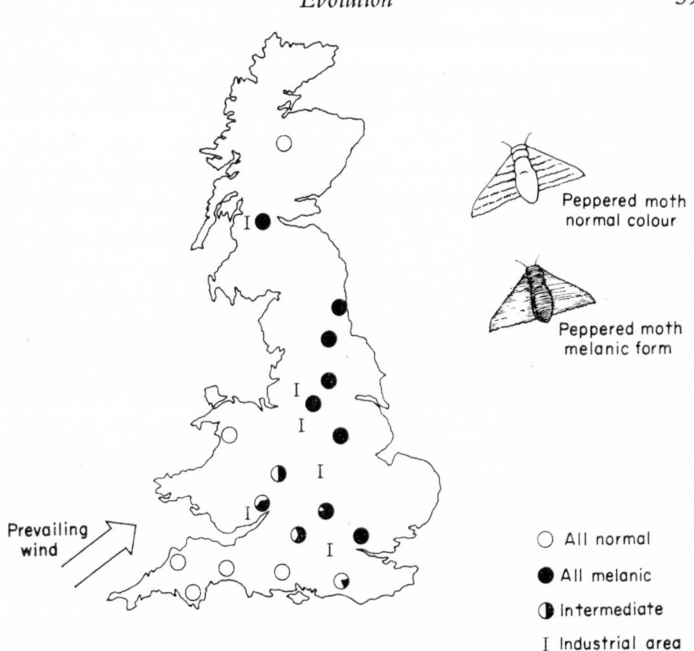

FIG. 3.5. The correlation between the number of melanic forms of the peppered moth and the degree of industrial blackening of the environment was established by H. B. Kettlewell.

Some examples of the operation of selection in nature that have been subjected to exact analysis are those of Cain and Sheppard on the colours of banded snails, and Kettlewell on industrial melanism (Fig. 3.5), although there is much other work on semi-artificial populations such as *Drosophila* in culture and general studies on selection of resistance in bacteria and insects to poisons. One of the most recent of these researches is also the most interesting, and it concerns the spread of dark or melanic forms of moths in England after the blackening of the country by industrial deposits some time around the mid-nineteenth century. Early records for certain varieties of moths, particularly the peppered moth, *Biston betularia*, show that in the north the dark form was very rare prior to 1848, occurring at less than 1 per cent of the population. By 1900 the dark form was completely dominant as it was in other parts of the country blackened by soot from towns and factories.

Kettlewell actually transfered dark moths to woods in the south and light ones to the north, and by releasing and recapturing the moths was able to work out the actual rate of selection against conspicuously coloured moths. He found that an advantage of some 14 per cent was conferred on moths of the correct colour to blend with the environment—a very large selective advantage indeed and more than sufficient to change the population in a short period.

Neo-Darwinism

The modern synthesis of genetics and Darwinian selection is called neo-Darwinism. It postulates that any population consists of many individuals with slight differences of genetic constitution but which can potentially interbreed. Under the process of natural selection, gene complexes well suited to the environment will be preserved while those less favoured will not. At the geographical

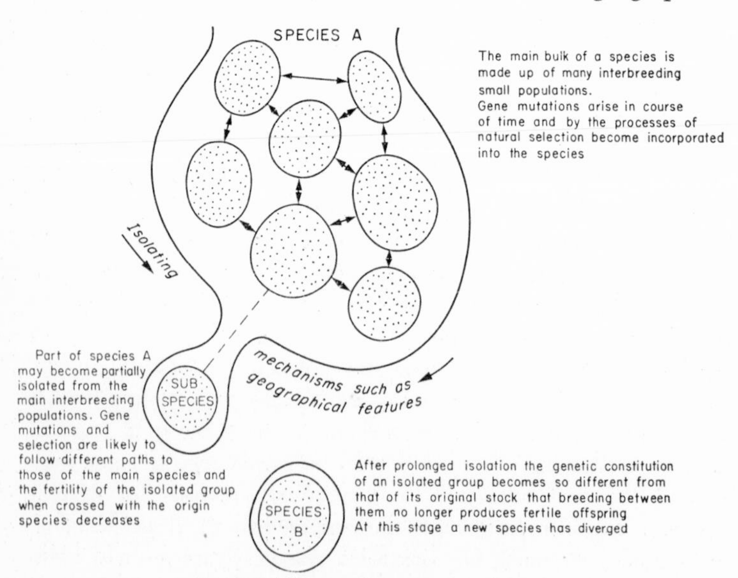

FIG. 3.6. The mechanism of evolution.

fringes of the population, or wherever small units have become isolated or semi-isolated by geographical or other conditions, the process of selection will be different. This means that gene complexes will arise substantially different from those of the main interbreeding body. Should the conditions causing isolation continue for a sufficiently long period—as may happen to a small unit cut off on an island—the divergent gene complex will become progressively less fertile when crossed with the gene complex of the original population from which it was derived. During the formation of sex cells it is necessary for homologous (one from each parent) chromosomes to line up together, and this can only be done successfully if the genes making up the chromosomes are sufficiently similar (Fig. 3.6).

At this stage the isolated unit may show all the characters of a sub-species and yet if it is rejoined to the parent species its new divergent characters may well be absorbed within the whole population. If, on the other hand, isolation continues, a new species is likely to be formed fully fertile only among itself and in some way more adapted to its particular environment than its predecessors.

There are many examples of this process of specialization going on in our own time—the wren of the mainland of Great Britain has divergent sub-species in Iceland, the Faroes, Shetland, and St. Kilda—sub-species which are likely to diverge more and more widely as time goes on until complete genetic isolation is reached between them.

Neo-Darwinism explains evolutionary divergence in this sort of way, and it is a combination of the new systematics and genetics described in other chapters with the important element of competition and selection demonstrated a century ago by Darwin and Wallace.

Questions on Chapter 3

1. In what ways did the theories of Lamarck about the mechanism of evolution differ from those of Darwin?
2. Outline briefly the main classes of evidence that support the belief that organic evolution has occurred.
3. What is meant by Neo-Darwinism? Give an account of any recent work on the operation of natural selection.

CHAPTER 4

Understanding the Mechanism of Heredity

IT IS clear that the offspring of two individuals both resemble and differ from their parents. The study of the way in which these differences and similarities arise and are passed on is called heredity, and it has been a matter of speculation to the human race from the earliest times.

Understanding heredity is not of mere academic interest. This science of character transmission, which has come to be called genetics, enables effective breeding programmes to select those specific qualities we require in our domestic animals and our crops. Such selective breeding has always been practised by man (Genesis 30.37) but it becomes very much more effective once the underlying principles are understood.

Changes in the heredity factors, or genes, which are termed mutations, yield important information about evolutionary mechanisms and how living organisms came to diverge into a multitude of varying species.

As far as the human race itself is concerned, genetics are important both socially and medically. The application of genetic knowledge to human reproductive habits is called eugenics, but such application is rightly of limited use in a free society. While it is desirable that the best or "fittest" members of a community should be the parents of the next generation and that those with serious heredity defects should refrain from passing them on, such an ideal is best obtained through the spread of knowledge and voluntary choice.

A final importance of heredity and mutation has come with the advent of nuclear power and its possible consequences on the load of harmful genes carried by man.

Early Ideas on Heredity

Much of the results of modern research into the mechanism of heredity comes from the application of exact analysis to controlled breeding experiments as well as microscopic study of the heredity matter. Such techniques were not available or were not used by early investigators, and much that they thought about heredity was either speculation or guesswork.

As we have seen to be true (p. 2) of many biological notions, definite ideas were put forward by the Greek philosophers. Some supposed the male to contribute a small animalicule which would grow in the body of the female, but whether this was true, or whether male and female contributed equally to the next generation, was not really known. Aristotle held the male influence to be predominant, while the Epicurean school held the more modern view of an equal participation in determining the characters of the offspring.

These and other less sensible views were handed down as far as the seventeenth and eighteenth century more or less unchanged when drawings of the sperm began to be made using the poor quality microscopes of the times. Some of these drawings show the sperm as a tiny man or homunculus (Fig. 4.1), and similar tiny

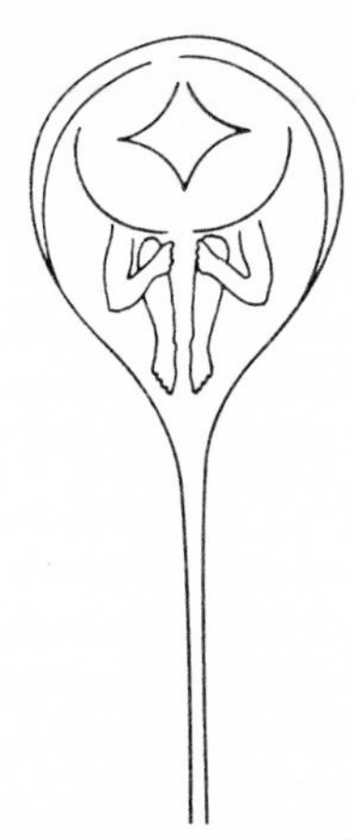

FIG. 4.1. Early representation of human sperm.

people were discerned in the egg the discovery of which, for the mammal, was made by de Graaf (1641–73). These false pictures of the sex cells made people believe that the sperm and egg themselves carried the seeds of the next generation in their minute bodies and this seed, itself a tiny homunculus, carried its seed also and so on, like a series of wooden Russian dolls or Chinese boxes. Thus it was supposed that Adam or Eve, according to which sex was favoured, had the whole future human race within their sex cells.

Writing in the mid-nineteenth century, Charles Darwin's speculations on heredity were but little in advance of the preceding ideas and are the least satisfactory of all his biological works. Darwin thought that all the organs of the body make tiny hereditary particles which become transmitted to the germ cells and incorporated into the material of the sperm or egg. This process he thought to happen throughout the lifetime of the individual. Basically these ideas came from the Hippocratic school of ancient Greece and are known as pangenesis. They seem to be without any foundation and there is no evidence of the transfer of normal genetic material from the body to the sex cells.

Much nearer to an understanding of heredity was Darwin's cousin (the whole family seems to have had a remarkable talent for scientific investigation) Galton, who was writing in the latter half of the last century. Galton had tried to apply quantitative methods to his study of heredity, and for man he showed that such characters as height, intelligence, fertility, the formation of mental images, and eye colour are handed down quantitatively. He suggested that the heredity makeup of an individual is compounded a half from his two parents, a quarter from his four grandparents, an eighth from his eight great-grandparents, and so on to the full total. This seems to be a logical idea, but his methods of assessing many characters at once, as well as dealing with characters which are determined by many individual genes (see p. 51), led him to confusion and obscured the simple quantitative mechanism of character transmission which was made so clear by Mendel. The latter was more or less contemporary with Galton's, but his work was not given publicity until the beginning of the twentieth century, long after his death.

The First Discoveries of Genetic Material within the Sex Cells and Research into the Nature of Genetic Variation

It has already been made clear that heredity operates and is transmitted at the microscopic level of the germ cells while the manifestation of inherited characters is seen in the whole macroscopic organism. The discoveries in this field of biology also came at these two levels, and it is only very recently that we are beginning to knit up the cytological event with its manifestation in the organism.

In 1879 the Swiss Fol saw the sperm entering the egg to bring about fertilization, and in the same year the chromosomes and the process of their division, or mitosis, was described. Important in this field of cyto-genetics were the discoveries and ideas of August Weismann (1834–1914).

Weismann understood that the body and the sex cells were in a sense isolated from each other in regard to their heredity material. The body, or soma as it is called biologically, is a sort of "house" constructed by the original first cell and in which the germ or reproductive cells can live. While the body will die in due course, the germ cells are passed on to make the next generation and are, in this sense, immortal (Fig. 4.2). This concept was called the "continuity of the germplasm" and it was coupled with the conclusion that changes made during life to the body cells would not affect the germ cells, hence the next generation. Such a conclusion was obviously opposed to the inheritance of acquired character doctrine of Lamarck.

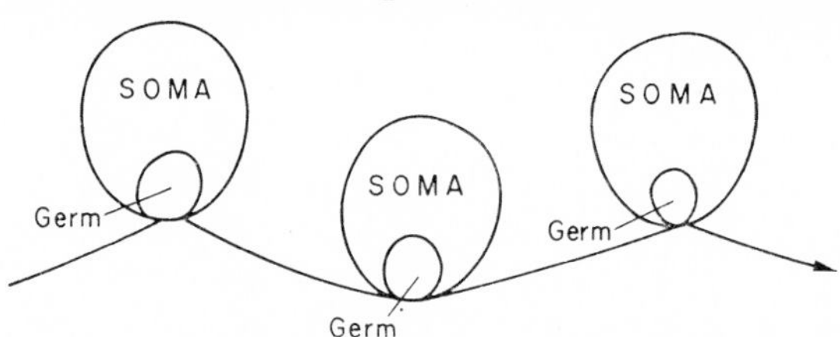

FIG. 4.2. Weismann showed the separation of the germ cells from the rest of the body and the continuity of the former from one generation to another.

Besides these important ideas, Weismann suggested that the heredity characters, or genes as we call them now, were carried on the chromosomes and that some sort of halving of genetic material must take place before sexual reproduction took place. This latter was a brilliant deduction, as meiosis or reduction division was not actually described until 1889 by van Beneden.

While the work of Weismann was well known to his contemporaries, quite otherwise was that of Gregor Mendel, whose paper on "Experiments in plant hybridization" was published in a somewhat obscure local natural history journal in Brunn in 1866 and 1869. The experiments that Mendel carried out in the gardens of his Austrian monastery were not to influence genetic thinking for a further 35 years.*

Random variation among the offspring of two individuals has been a key part of Darwin's evolutionary theory (see p. 31) and a good deal of investigation was made into such variations by those interested in heredity. One of the first workers on "Discontinuous variation" was Bateson, who published a book of experiments and suitable materials for study in 1894. Shortly afterwards the Dutch botanist de Vries began his investigation into the heredity of the evening primrose, *Oenothera*. These two pieces of work showed how variations could appear between generations, and to such variations de Vries gave the name mutation. In fact it is now clear that many of these so-called mutations were only rearrangements of genetic material already present. In its strict sense a mutation involves an inheritable chemical change in the nature of a gene.

The Laws of Heredity—The Work of Gregor Mendel

With this new interest in the results of quantitative breeding, de Vries and other workers began to look through the earlier literature and soon Mendel's paper was rediscovered by three independent researchers. It was obvious that here was a new and exciting breakthrough in the understanding of heredity that was to influence all genetic thought.

---

*In fact, Mendel's paper was sent to scientific societies in Germany and to the Royal Society of England. Despite its clarity and importance, its significance was just not recognized at the time.

Using the pea plant, *Pisum*, Mendel had selected to follow the heredity of certain contrasting characters that could be readily followed from one generation to the next. Among the pairs o characters that he followed were the tallness and shortness of the plant, the nature of its seed coat wall, i.e. whether it was round or wrinkled, and the colour of the cotyledons, namely yellow or green. Allowing self-fertilization to occur as it normally does in peas, he bred pure lines of plants having one or other of these characters and then made deliberate crosses between these lines and counted the types of offspring. He found that his contrasting characters (e.g. tallness and shortness) segregated out in definite proportions in the second filial* generation, and that one character seemed to dominate and mask the other when both were present; i.e. the plants of the crosses were either all tall or all short.

The cross between tall and short lines can be illustrated as follows:

Parent (P)	Tall		×	Short
F_1			Tall (selfed)	
F_2	Tall	Tall	Tall	Short

If we express this using modern symbols of *T* for the tall dominant character and *t* for the short recessive one and include the heredity material of the gametes (which are formed from halving of the parental gene material) the result is:

P	*TT*		×		*tt*	
Gametes (G)	*T* *T*				*t* *t*	
F_1		*Tt*		*Tt*		(all tall)
G		*T* *t*		*T* *t*		
F_2	*TT*	*Tt*		*Tt*	*tt*	
		or 3 tall, 1 short				

*The first filial generation are the offspring of a cross and the second filial generation are the second generation from the original cross.

Mendel expressed these results in his first law which is: *When two pure bred individuals, having contrasted characters, are crossed the characters segregate out in definite proportions in the second filial generation.*

These experiments with one pair of characters were then followed up with breeding and crossing pure lines of peas with two or more pairs of contrasting characters. To take an example: tall peas with yellow cotyledons may be crossed with short plants with green cotyledons. The results of such a cross are:

P	tall yellow	×	short green	
F_1	tall yellow (selfed)			
F_2	9 tall yellow	3 tall green	3 short yellow	1 short green

or, using the symbols *T* (tall), *t* (short), *Y* (yellow), *y* (green):

P	*TTYY*	×	*ttyy*	
G	*TY*	*TY*	*ty*	*ty*
F_1	*TYty*	*TYty*	(all tall yellow)	

Because the two pairs of characters segregate out quite independently the gametes or sex cells from the F_1 are *TY*, *Ty*, *tY* and *ty*, and by putting them in the form of a table (Table 1) which shows all possible mating combinations, Mendel's results are obtained.

TABLE 1

Ovules \ Pollen	*TY*	*Ty*	*tY*	*ty*
TY	*TYTY* tall yellow	*TYTy* tall yellow	*TYty* tall yellow	*TYty* tall yellow
Ty	*TyTY* tall yellow	*TyTy* tall green	*TytY* tall yellow	*Tyty* tall green
tY	*tYTY* tall yellow	*tYTy* tall yellow	*tYtY* short yellow	*tYty* short yellow
ty	*tyTY* tall yellow	*tyTy* tall green	*tytY* short yellow	*tyty* short green

That is, 9 tall yellow, 3 tall green, 3 short yellow, 1 short green.

These results are expressed in Mendel's second law: *When two pure bred individuals showing two, or more, pairs of contrasting characters are crossed, the characters segregate out independently in the second filial generation.*

The real significance of these laws and experiments are that they show that the heredity material exists in units, or genes as they are now named, which do not blend but which segregate out according to simple quantitative ratios. For this reason Mendel's theories are classed as those of "particulate inheritance".

We realize from subsequent work that Mendel was either lucky or skilful in choosing the characters that he did and that most characters are not determined by a single unit or gene. This means that while most do not segregate out to give the simple ratios described above, the germinal unit is still the basic entity involved and Mendelian segregation, though often obscured, still operates.

The Development of Mendelism

In the first two decades of the present century genetic discoveries came rapidly. Sutton was able to demonstrate genetic interchange between parental chromosomes during reduction division and suggested how a great variety of dominant and recessive characters could become grouped together in an offspring. These studies of Sutton were among the first in the science we call "cyto-genetics" where the results of breeding experiments can be correlated with events taking place within the reproductive cells. It soon became clear that there must be many more heredity units, or genes, than there were chromosomes in an individual, so that many genes are linked together on a single chromosome thread. This idea of linkage was confirmed experimentally by Bateson and Punnet in a series of papers from 1905 onwards. Linked genes do not behave in a Mendelian fashion as they cannot segregate independently. Thus Bateson found that in the primula (*Primula sinensis*) the length of the style and the flower colour were linked. On breeding a pure bred

short style × long style
blue flower   red flower

the F_1 were all short style blue flower as these genes are dominant. When this F_1

was selfed on Mendel's second law it would have been expected that four types of plants would occur in the ratio 9 short style, blue flower, 3 long style, blue flower, 3 short style, red flower, 1 long style, red flower. In fact the two characters do not segregate independently as they are on the same chromosome, and the actual result was 3 short style, blue flower: 1 long style, red flower in the F_2. Linked genes can be represented in □ 's or placed above each other ie. $\frac{S}{B}\frac{S}{B}$ and $\frac{s}{b}\frac{s}{b}$ rather than *SSBB* and *ssbb*.

We can now set out the cross as follows:

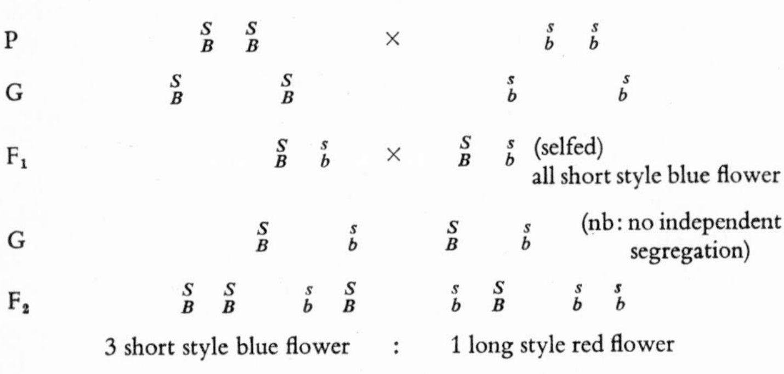

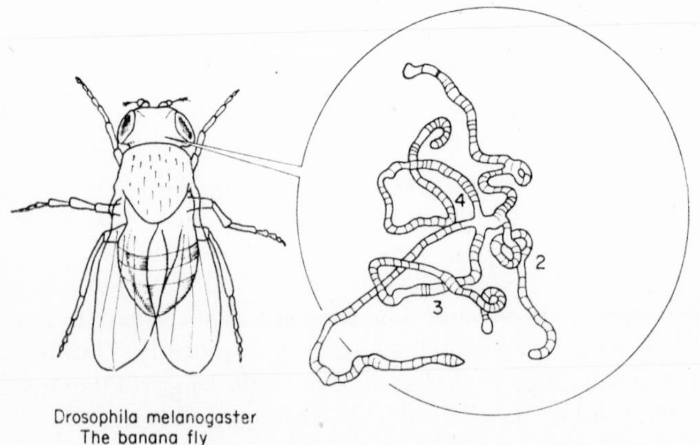

FIG. 4.3. The giant chromosomes from the salivary glands of the fly. It has been possible to map the relative positions of many genes along the chromosomes.

Besides their work on linkage, Bateson and Punnet also showed that many characters, such as the colours of sweet peas and combs of poultry, must be due to the interaction of several genes and that the one-gene-one-character of Mendel was fairly rare.

In 1910 the American geneticist T. H. Morgan produced the first work on the banana fly *Drosophila melanogaster*, on which so much experimental study has since been made (Fig. 4.3). This small insect was easy to culture and bred rapidly and, as it also turned out, to have giant salivary chromosomes, so in every respect it was an ideal object for genetic investigations. Morgan soon showed mutant genes, such as white eye, and soon another worker, Sturtevant, found a method of plotting the distribution of the genes along a chromosome. The basis of this method is that linked genes have a

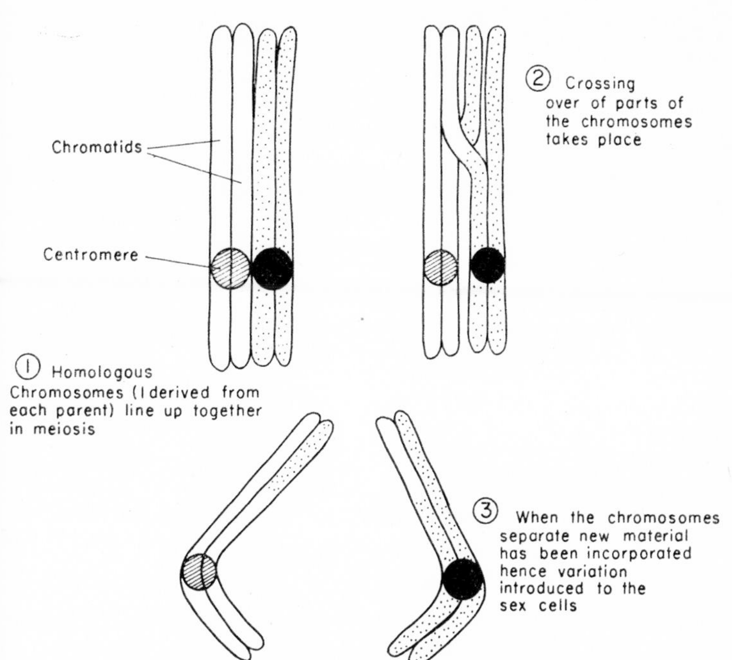

FIG. 4.4. The crossing over of parental chromosomes at formation of sex cells.

certain frequency of cross over at meiosis, and the results of cross overs can clearly be detected in the offspring (Fig. 4.4). Genes that are close together on a chromosome are less likely to cross over. The frequency of cross over between linked genes can be used as a measurement of the distance between them, and these units were called "Morgans". By sufficient crosses it was possible to plot the whole chromosome maps of *Drosophila* and to show that the number of linkage groups in fact corresponded to the number of chromosomes (Fig. 4.5).

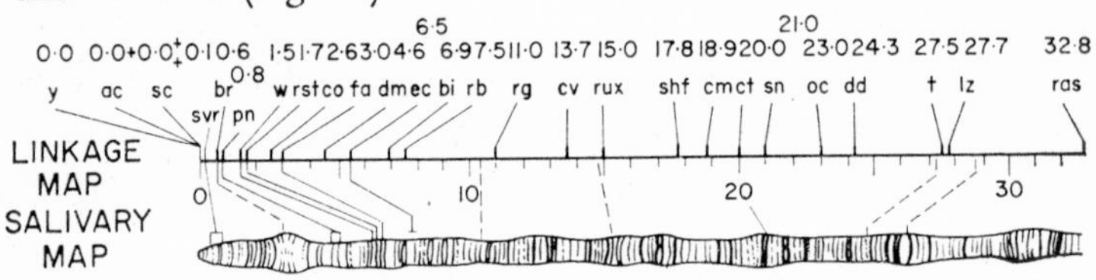

FIG. 4.5. The genes on a single *Drosophila* chromosome.

An example of the cross over of linked genes in *Drosophila* is given below.

Drosophila. Aristopedia antennae and ebony body are linked and are recessive to normal antennae and grey body.

Normal antennae Grey body × Aristopedia antennae Ebony body

N N / *G G* × *n n* / *g g*

all Normal antennae Grey body (*N n* / *G g*) back crossed with Aristopedia antennae ebony body (*n n* / *g g*)

N n / *G g* × *n n* / *g g*

Normal antennae Grey body	Normal antennae Ebony body	Aristopedia antennae Grey body	Aristopedia antennae Ebony body
44%	6%	6%	44%
N n / *G g*	*N n* / *g g*	*n n* / *G g*	*n n* / *g g*

12% cross over classes

Therefore the genes Aristopedia/Normal } antennae and { Ebony/Grey } body are 12 MORGANS apart on their chromosomes.

Further Discoveries on the Nature of Variation

While Mendel's laws were being tested and elaborated in the first decade of the century, Johannsen pointed out that the sort of results that were appearing would not actually account for evolution. Cross over and assortment of a fixed number of variables does not really introduce anything new any more than the shuffling of fifty-two playing cards can produce more than a certain number of hands, and there was a great deal of confusion about the whole nature of variation. Indeed, to many it seemed that there was insubstantial basis for belief in Darwinian selection bringing about speciation.

It was not until 1927 that the work of H. J. Müller made clear the true origins of evolutionary change and vindicated Darwin. Müller had subjected a *Drosophila* culture to X-rays and found that the radiation caused a great number of new variations or mutations to appear. These were proper mutations in the sense that they were inheritable chemical changes. The interesting thing about Müller's results was that a very high percentage of these mutations were harmful or actually lethal. It appeared, therefore, and it still does, that the basis of evolutionary change is the tiny number of beneficial mutations that arise in a species.

One of the causes of variation was certainly radiation, but mutations also seem to arise spontaneously by other means. Certainly the modern theory of evolution, often called neo-Darwinism, supposes a supply of mutations with the beneficial ones incorporated into a population by the processes of natural selection.

The Work of the Gene in the Cell

In the last 30 years much evidence has accumulated about the role of the gene in determining the nature of the cell and how the gene carries its information. As will be remembered, the complexity of gene action and interaction had been suggested by the early work of Bateson and Punnet. The next major new work in this particular field came out in 1941 and was done on the mould *Neurospora* by Beadle and Tatum (Fig. 4.6). Irradiated spores of the mould were found to give rise to individual plants that could not synthesize certain of the essential growth substances that the normal plant was

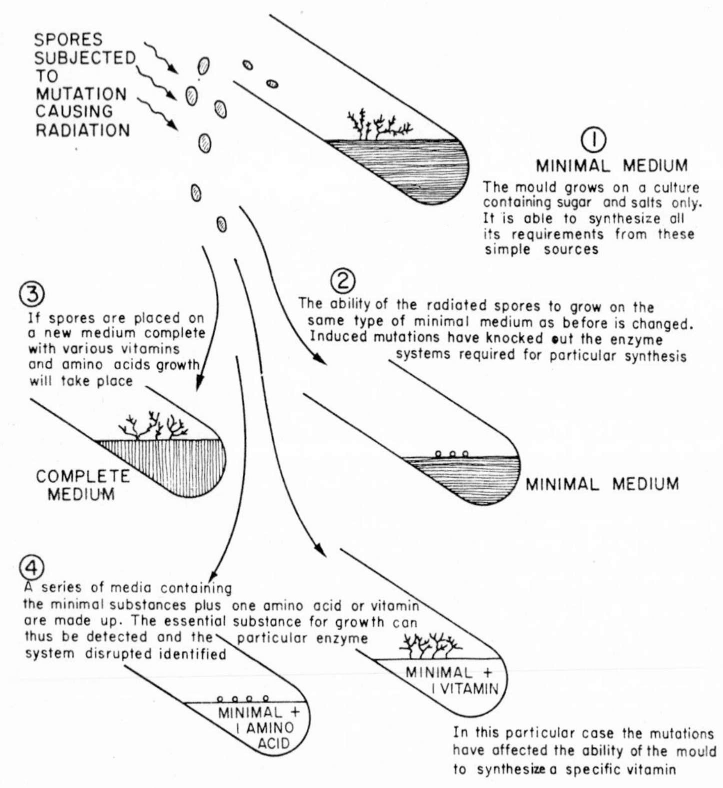

FIG. 4.6. The connection between genes and enzymes was established by work on the fungus *Neurospora*.

able to make. They showed this by getting the irradiated and mutated moulds to grow only after the addition of specific substances. The conclusion of this work was that individual genes were concerned in the manufacture of actual enzymes in the cell.

Recent investigation has been into the actual chemistry of this genetic material and what connection it had with enzyme synthesis. An important paper was published by Avery and his co-workers in 1944 who isolated a chemical substance from a capsule-forming species of pneumonia bacteria that could induce capsule form-

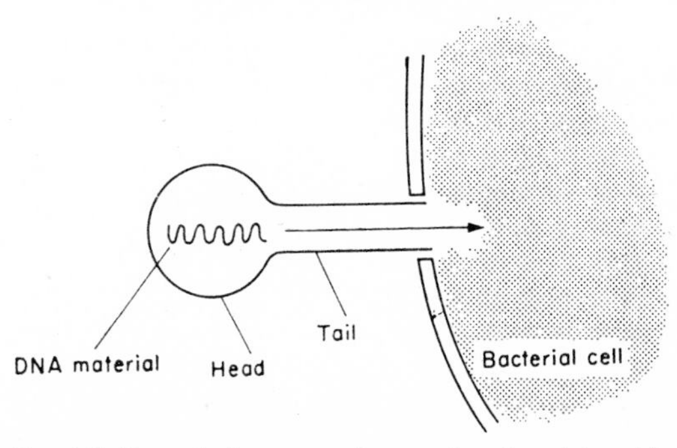

FIG. 4.7. Bacteriophages consist mostly of nucleic acid.

ation in other strains. Such material obviously had the necessary characteristic properties required of a gene, and on analysis these workers found it to be the substance desoxyribose nucleic acid* (more commonly known as DNA). By the early 1950's Lederberg showed that the bacteriophage, a microscopic organism that attacks bacteria, consisted almost entirely of DNA and that it was capable of transmitting genetic information from one bacterium to another (Fig. 4.7).

By this time it was established that the gene was a region of this desoxyribose nucleic acid molecule, and it only remained to find its structure and its code. A start on this was made by the brilliant studies of J. D. Watson and F. H. C. Crick who in 1953 submitted their model of the DNA molecule. As is now generally known, the Watson–Crick model consisted of a double helix whose spirals were made of alternate phosphates and ribose sugars and whose arms were linked together by pairs of bases—adenine with tyrosine and cytosine with guanine (Fig. 4.8). The arrangement of these pairs of bases was taken to be the means of coding chemical information about the relative positions of amino acids in the making up of the cells enzymes. A code which used three pairs of non-overlapping groups would be sufficient to code the twenty-odd amino acids involved in all protoplasm.

This triple code has been tested by adding or subtracting bases

*Or deoxyribose.

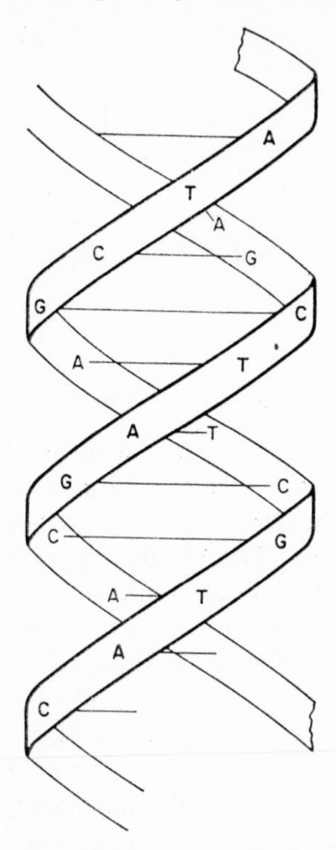

FIG. 4.8. The gene molecule.

from a DNA strand. If one or two base pairs are inserted or removed, the strand reads nonsense, i.e. it does not make a sequence template. On the other hand, if three base pairs are inserted or removed the positions subsequent to this interference work as coding units.

Within the last few years the relationship between the gene material in the nucleus and the related ribose nucleic acid in the cytoplasm has become clear (Fig. 4.9). Messenger RNA carries positional information continuously out from the nucleus to the RNA microsomes in the cytoplasm where actual amino acids are condensed into proteins. Even the specific base code corresponding to individual amino acids is being determined.

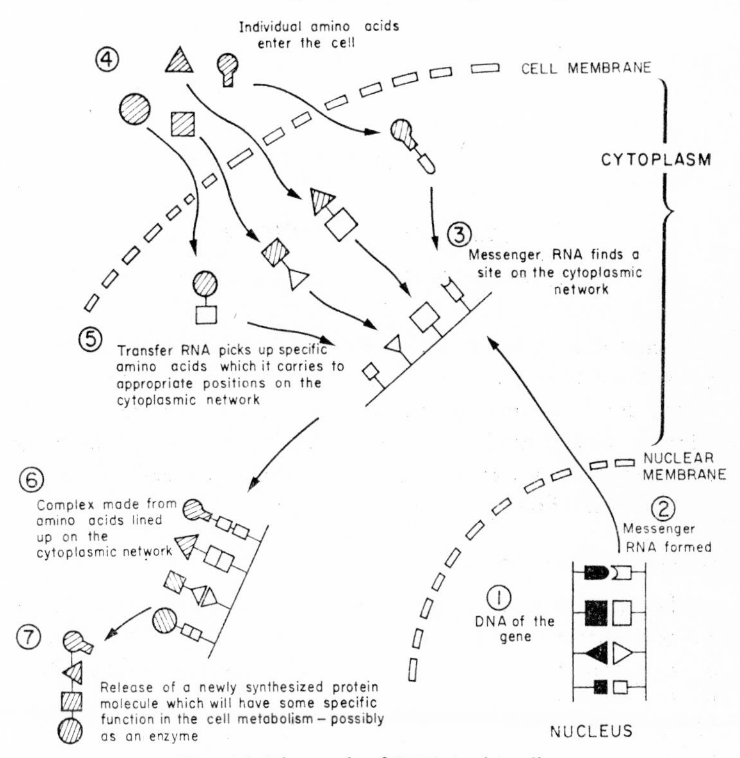

FIG. 4.9. The work of DNA in the cell.

As a result of this progress—and other work on pre-natal immunization—it may be that we shall eventually be able to correct inborn errors of metabolism such as certain forms of idiocy and diabetes.

At present this day is still a long way off.

Questions on Chapter 4

1. Write shortly on the gene code, Weismann, mutations.
2. Discuss the importance of Mendel's work in the development of our knowledge of heredity.

CHAPTER 5

Aspects of Medicine

Part I: *The Growth of Anatomy, Physiology and Surgery*

THE contributions made by biology to medicine cover so much ground and so many scientific disciplines that it has been thought best to divide the material into two parts. In the first we shall consider the way in which an understanding of the parts of the body and their functions arose, and in the second we shall follow the establishment of the germ theory of disease.

Medicine before the Greeks

The beginnings of medicine are quite lost in antiquity, but it seems clear that each of the great civilizations that preceded or were contemporaneous with the Greeks had developed some form of the healing of wounds and the treatment of disease. In every case the art of healing was bound up with the mysteries of religion and priesthood, and this is not surprising as it was generally believed that diseases were due not to natural causes but to the malign influences of gods or evil spirits.

From China we find a collection of medical writings dating from the third millennium B.C. Although Chinese medicine seems to have been fairly advanced in some respects (e.g. a crude form of immunization against smallpox existed), its great weakness was its lack of surgery. The religion of ancestor worship inculcated a reverence for the dead; little dissection was carried out, so that knowledge of anatomy was necessarily sparse. The main physiological idea underlying the treatment of disease was that the body was animated by a Tao or natural spirit which existed as two components —Yin and Yang. When these two entities got out of balance, disease resulted, and such practices as acupuncture (the sticking in of needles to various parts of the body) were aimed at redressing the balance. There is a parallel here to the Greek and later notions of humeral imbalance.

Indian medicine was highly sophisticated, and in writings which date from B.C. but which incorporate a very much earlier oral tradition, we find a complex account of disease, diagnosis, prescription, and medical instruments. There was an accurate knowledge of anatomy, and literally hundreds of bones and muscles were described. Of all the early civilizations the Indians used the most extensive range of medicinal plants. The very first hospitals in the world were set up by ancient kings of India.

As with Chinese medicine, we find Egyptian medical writings of great antiquity (2500–2000 B.C.) in which various diseases are described together with remedies. There are also accounts of the treatment of fractures. Some of the Egyptian "medicines" include faecal matter disparagingly referred to as "sewage pharmacology" until a few years ago, when it was discovered that faeces and soil are rich in antibiotic substances. Perhaps these despised medications really did have some healing properties. The Egyptians were experts at the embalming of bodies, and this automatically led to a fair knowledge of anatomy. They thought that essences of different sorts were driven to the parts of the body by the heart through tubes, and they understood the importance of the pulse as an aid in diagnosis.

By means of trade routes, exchanges of ideas between Egyptian and Indian medicine took place.

Greek Medicine

The Greeks were at least partially successful in separating medicine from religion and magic and putting the subject on a clinical basis. Although their knowledge as to the causes of disease were mostly erroneous, this did not detract from their methods of accurate observation and recording of symptoms.

In the third century B.C. Empedocles suggested, or at least crystalized, the idea that the body was composed of equal parts of earth, air, fire, and water. This notion, together with the theory that good health resulted from a correct balance of these elements, was incorporated in the writings of the Hippocratic school. The elements, or humours as they were called, became equated to various body substances and dispositions. Thus the earth was equivalent to black bile (or choler), and when dominant gave the individual a sad and melancholic (literally black bile) nature. The air was yellow bile and led in excess to a bad tempered or choleric personality. Fire was equal to the blood and produced a happy or sanguine disposition, and finally water was the phlegm and made for a cold or phlegmatic nature. These terms are still used to describe personalities.

Whether Hippocrates himself, often described as the Father of Medicine, actually wrote any medical texts, seems open to doubt; but a large number of works, extending over several centuries, was put together and became known as the Hippocratic collection. It is in these writings that we find the famous Hippocratic oath which lays down the highest ethical standards for the practitioner. Thus:

> The regimen I adopt shall be for the benefit of the patients according to my ability and judgement and not for their hurt or wrong. I will give no deadly drug to any though it be asked of me nor will I council such and especially I will not aid a woman to procure abortion. Whatsoever things I see or hear concerning the life of men in my attendence on the sick or even apart therefrom, which ought not to be noised abroad, I will keep silence thereon, counting such things to be as sacred secrets. Pure and holy will I keep my life and my art.

Accurate clinical observation led to the grouping of symptoms and the recognition of specific diseases whereby a prognosis, or foretelling, of the course of a disease could be made. Among others, mumps was recognized. As far as actual treatment went, the armoury of the Greek physician was naturally rather limited. There were over 200 medicinal herbs in use (see p. 16), and much use was also made of regulating the diet. The letting of blood, so much used in later centuries, was little practised.

Among the Hippocratic collection are several works dealing with the treatment of fractures and dislocations and which show accurate knowledge of the anatomy and functioning of bones and joints. The use of splints to keep the limb rigid, and to a lesser extent the use of traction (so it does not set short) were understood. Neuro-surgery, if we call it that, was confined to boring holes in the skull, called trephining. This latter practice was to let out "evil spirits" and was paralleled in other ancient civilizations. The Greeks had some idea of the surgical treatment of teeth.

Typical of the Hippocratic approach is this account made of the sacred disease, supposed at the time to be due to the ill-wishing of the gods.

> Those who talk about such diseases as being due to the Gods, in fact, treat the disease with all sorts of incantations and magic, but they are also very careful in regulating the diet. Now if food makes the disease better or worse how can they say it is the Gods that do this? The fact is that the invoking of the Gods to explain diseases and other natural events is all nonsense. In nature all things are alike in that they can be traced to preceding causes.

This sort of empirical and reasonable approach yielded practical advances, and it shows why Greek medicine was head and shoulders above that of earlier cultures.

Alexandrian Medicine

With the breaking up of the Greek Empire the main centre of medical progress moved to Alexandria in Egypt where, in the third century B.C. a school of medicine flourished. Among its members were two Greek physicians Herophilus and Erasistratus, and mainly to them are attributed several important anatomical and physiological discoveries.

The Egyptians themselves had a sound knowledge of human anatomy derived from the preparation of bodies for embalming, an art at which they excelled. In the Alexandrian school dissections were made and the structural differences between arteries and veins described. The brain was recognized as the seat of intelligence and sensation, and nerves associated either with motion or feeling. It was thought that the body was worked by a spirit that was in some way related to the process of breathing. Movement was brought about by the inflation of the muscles by this spirit which was driven into them through the nerves.

It was held that diseases were due to an excess of blood, and therapeutic measures included dieting and bathing and exercise and, to a lesser extent the letting of blood.

Roman Medicine

The cultural influence of the Greeks is to be clearly seen on their less creative Roman neighbours, and this Greek influence was very marked in the development of medicine throughout the whole period of the Empire.

The first school of Roman medicine was run by the Greek physician Asclepiades (*c.* 40 B.C.) who is remembered for his idea that the health of the body can be maintained by the opening or shutting of its pores suitably induced by treatment. After this a number of state medical schools grew up in Italy, Gaul, and Spain, the main object of such schools being the training of army doctors and surgeons. The great herbalist Dioscorides (see p. 16) was a member of such an establishment.

The first Latin medical text was that of Celsus (A.D. 30), and it is written in the clear style of the earlier Greek manuscripts. Full descriptions are given of inflammations and the suppurations and gangrene which follow wounding. Operations on the removal of the tonsils, remarkably similar in technique to our modern methods, are described as well as skilled instructions for the amputation of limbs which involves the tying off of main vessels. The whole art of surgery as found in the writings of Celsus appears to be much more advanced from the Hippocratic or Alexandrian standards.

By far the greatest of the Roman physicians was Galen (A.D. 130–200), and his work and theories were to influence medicine until the new advances of the fifteenth, sixteenth, and more particularly the seventeenth centuries. Galen dissected the bodies of cattle, pigs, and Barbary apes, and wrote voluminously on structures and function, referring most of his discoveries to man. Although not actually a Christian, he approved of that religion and believed that every structure of the body had a definite function bestowed on it by a benign creator. This idea of design was to appeal to the ecclesiastical authorities of the Christian and, later, the Islamic religions, and in both we find Galen's work treated as being authoritarian.

As far as his anatomical descriptions are concerned, Galen was at his best when dealing with bone and muscles and at his worst when it came to the blood system (Fig. 5.1) and the nervous system. In trying to explain functioning of the body he suggested that the essential "life force" (or pneuma) was breathed in through the trachea and passed into the lungs and thence via the pulmonary vein to the left side of the heart. He believed that the life force then came into contact with the blood, itself manufactured from food in the

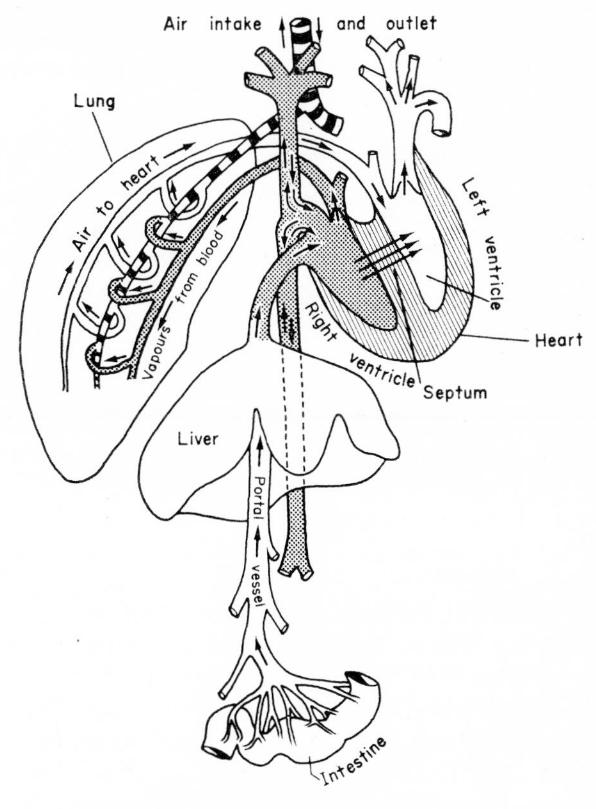

FIG. 5.1. Galen's circulation of the blood.

liver via holes in the septum which separates left and right ventricles. (This idea, i.e. the passage of blood from one side to the other was to persist for hundreds of years despite the failure to find the necessary holes.)

The blood now in the left ventricle was considered to be in the form of impure "natural spirit" and was transported to the lungs where it gave off wastes. While the impure blood was thought to ebb and flow in the veins the pure blood did the same in the arteries, and some of this latter, passing to the brain, gathered the most noble of all the essences the "animal spirit". It was this special blood flowing down the nerves of the body that endowed it with movement and sensation.

It was not until 1642 that Harvey refuted Galen's ideas and demonstrated the circulation of the blood.

Medicine in the Middle Ages

With the collapse of Roman civilization in the early part of the fifth century the study of medicine, and indeed of all natural knowledge, declined in the West. In Asia and Arabia, however, the old classical learning continued to be studied, but it was not to become generally available until the time of the Renaissance in the fifteenth century.

Western Europe meanwhile plunged into the period of its history known as the "Dark Ages" (but see also p. 2), and little progress was made in medical knowledge for some 500 years. The human body itself was considered sacred and dissection prohibited, so that fundamental advances in anatomy could not be made. Such learning that did survive was largely confined to the monasteries, but it was a very meagre thread and diseases, reflecting divine retribution, were thought to be best cured by prayer and invocation of the appropriate Saints. Surgery was forbidden to priests and fell into the hands of barbers and ignorant men who supposed that the formation of pus was a necessary step in wound healing and who applied to their wretched patients various foul and insanitary dressings.

By the twelfth century a number of universities had become established such as Paris, Padua, Bologna, Montpellier, Oxford, and Cambridge. At first medicine, or rather anatomy and physiology, was treated as part of natural philosophy, but later separate faculties were formed. Thus by the thirteenth century we find human dissection being carried out at Bologna and other medical schools.

Renaissance Medicine

In 1453 the Turks sacked Constantinople and the migration of scholars into the West made available much of the earlier writings and culture so long lost or debased. These original versions of the Greek and Roman knowledge seem to have stimulated once again the whole spirit of curiosity and investigation. The explorers of the fifteenth century had greatly extended the boundaries of the known world, while in western Europe the Reformation of the Church had begun. Everywhere there seems to have been a new spirit of questioning established dogmas and re-examining natural phenomenon. The Renaissance, herald of our own modern age and great bound forward in human cultural evolution, had begun.

One of the most remarkable figures of the Renaissance, in fact the period is usually taken as starting from his birth, was Leonardo da Vinci (1452–1519). Among his many interests was the way in which the human body moved, and to this end he made many accurate drawings of bone and muscle structures. Besides this he made investigations into optics and the structure and functioning of the eye as well as contributions to the anatomy of the lungs, brain, and nerves. Following Leonardo was Versalius (1514–64) who demonstrated anatomy at Padua and Bologna. At first he was inspired by the writings of Galen, but as he became aware of their many inaccuracies he began a new investigation of his own. His great books on anatomy were called *De Humani Corporis Fabrica*, but despite their merit they were not recieved kindly by the ecclesiastical authorities, daring as they did to dispute much of Galen's descriptions. Despite his inability to find the holes in the septum of the ventricle, necessary to Galen's theory of circulation, Versalius did not question the former's interpretation of the path of the blood. In fact some 25 per cent of human hearts do have prominent veins in the septum, and Versalius was not certain that these were not in fact passages from one side of the heart to the other,—i.e. Galen's "pores".

It so happened that the publication of Versalius's anatomy coincided with the writings of Copernicus about his discovery of the earth's movement around the sun. While neither of these men suffered death for their beliefs, less fortunate was their contemporary Servetus who was burned partly for his heretical views that the blood did not move in the way described by Galen.

Despite the advances made, the practise of medicine and surgery throughout the period was still at a very primitive level, although in the work of the French army surgeon Paré we see some return to the hygienic and refined operating techniques of the ancients.

The next major advance took place in the field of physiology which was established in 1628 as an experimental discipline by the English physician William Harvey. Inspired by his anatomical teacher Fabricius of Padua, who was himself working on the valves of the veins, Harvey returned to England and carried out his researches into the circulation of the blood.

In the first place he showed the output of the heart greatly exceeded the amount of blood that could be made from the food (see Galen's theory, p. 63), and in the second he showed how blood always travelled away from the heart in the arteries and returned to it in the veins. He could find none of the pores in the septum of the heart stipulated by Galen and he suggests that far from ebbing and flowing the blood travels from the heart to the lung, back to the heart, and thence to the body. In the tissues of the body he said there must be connections between the arteries and veins so that the blood passing to the latter is transported back once more to the heart. (These connections, which we call capillaries, were not seen by Harvey, but some years later were found by the pioneer microscopists Leeuwenhoek and Malpighi.). The circulation of the blood is set out in Harvey's *Exercitatio Anatomica de Motu Cordis et Sanguinis in Animalibus*, which is a model of scientific method and clarity. It is interesting that no mention is made of animal or vital spirit, but that the heart is treated purely as a mechanical device through which blood fluid is pumped.

While Harvey was making his classic study of the circulation a fellow countryman called Thomas Sydenham, known as the "English

Hippocrates", was making a study of diseases. He grouped them together according to their symptoms and described plague, cholera, typhoid, typhus, and malaria (treating the latter with a crude extract of quinine). It occurred to him that each sort of disease had its own particular nature and possibly its own particular cause—obvious enough to us but at the time an adventurous breakaway from the notion of humeral unbalance. Despite his abilities as a clinician, Sydenham unfortunately ignored the idea of contagion in the spread of disease although such a concept had already been postulated.

Eighteenth Century Medicine

By the mid eighteenth century a whole lot of advances had been made in medical science, and we see the basis laid for a good deal of our modern knowledge. For the first time some sort of integration was made between the understanding of the structure and function of the body and its organs and the actual practice of medicine.

The micro-structure of tissues such as the skin and blood as well as organs such as the lung and kidney was described and simple

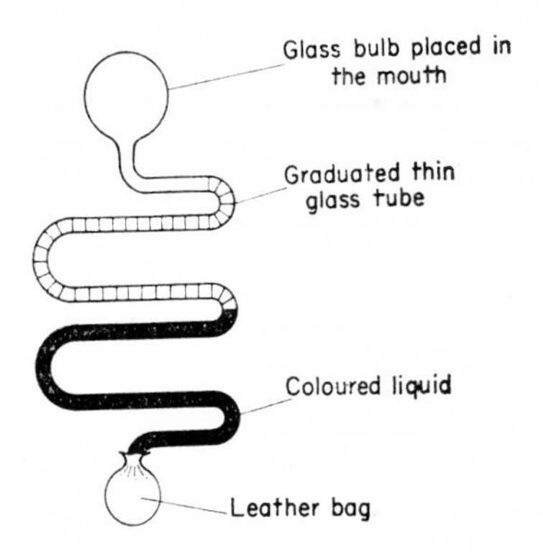

FIG. 5.2. The first sort of thermometer. The warmth of the mouth drove a quantity of the liquid out of a glass and into the leather bag. This quantity, measured by reading the scale of the tube, was proportional to the body temperature.

experiments made on the nature of the digestive processes. Notable was the work of Van Helmont (1577–1644) who interpreted the functions of the body as chemical processes. The taking of temperature (Fig. 5.2) as a means of diagnosis was known, as were the first early researches of Stephen Hales (see p. 102) on the pressure of the blood and its variations. While the results of mechanical investigations into function made philosophers such as Descartes believe it was only a type of complicated machine, there existed vitalists who maintained (and indeed continued to do so until organic syntheses were made in the mid-nineteenth century) that biological processes are not fully subject to mechanical analysis.

A correlation between the symptoms of disease observed during life and their relation to post-mortem examination was made by an Italian called Morgagni. From his work grew the idea that diseases may be associated with failure of specific organs, and anatomy and pathology became linked to give a clearer understanding of the whole nature of disease. It is only now that we find a drawing away from the classical notions of the humours and their balance as the factor determining health. Because of this new correlation, diagnostic techniques were developed for the examination of organs within the living patient. It was found, for example, by Avenbrugger, son of a brewer, that the noise made by striking two fingers against the chest varied according to the contents—air, solid tissue, or fluids all producing characteristic sound. Towards the end of the eighteenth century a form of stethoscope was introduced whereby the sounds of the heart could be heard.

In physiology the *Elements of the Physiology of the Human Body* by the Swiss Haller (1759) contained much new material on the working of the respiratory and digestive systems as well as some account of embryological development. The material of the brain and nervous system provided a basis for the discovery of the reflex arc early in the following century. Meanwhile the fundamental chemistry of the composition of air and of the products of combustion and of respiration led Lavoisier to the belief that respiration itself was a sort of "burning" of fuel that took place in the lungs. This concept is, of course, only partially correct, and it was to be another 100 years

The properties of the plum tre and hys fruite out of Dioscorides. 105

Prunus syluestris.

stop the belly. The gum of the plum tre gleweth together. If it be dronken with wyne/it breketh the stone and healeth the skurfenes of childer.

Out of Galene de simplicibus medicamentis.

He fruite of the plũ tre louseth the belly/ but more when as it is moyst & fresh/ & lesse when it is dry. But I can not tell what made Dioscorides to wryte yͤ dryed Damascene plumes do stop yͤ belly/ when as they do manifestly louse yͤ belly/but they yͭ cõ out of Spayne ar sweter. The trees answer in propor tiõ of qualite wᵗ yͤ fruites. The fruite of the wild plum tre is manyfestly byndyng and stoppeth the belly.

Out of Galene of the toures of norishmentes or meates

How shalt seldum fynde the plũb tarte or sour or to haue any unplesantnes/ whẽ it is

PLATE I A page from William Turner's *Herbal* (1568). Reference is made to the plum tree and its properties.

See Page 16

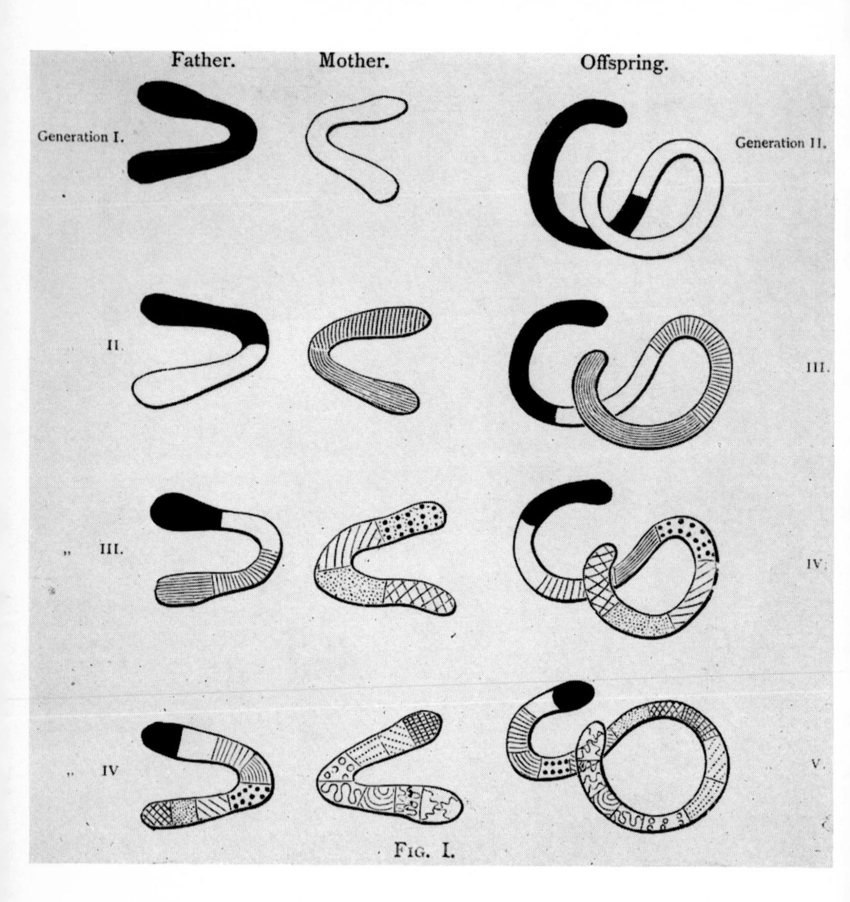

PLATE II August Weismann predicts meiosis and shows how the chromosomes might be supposed to carry heredity factors (1889).

See Page 45

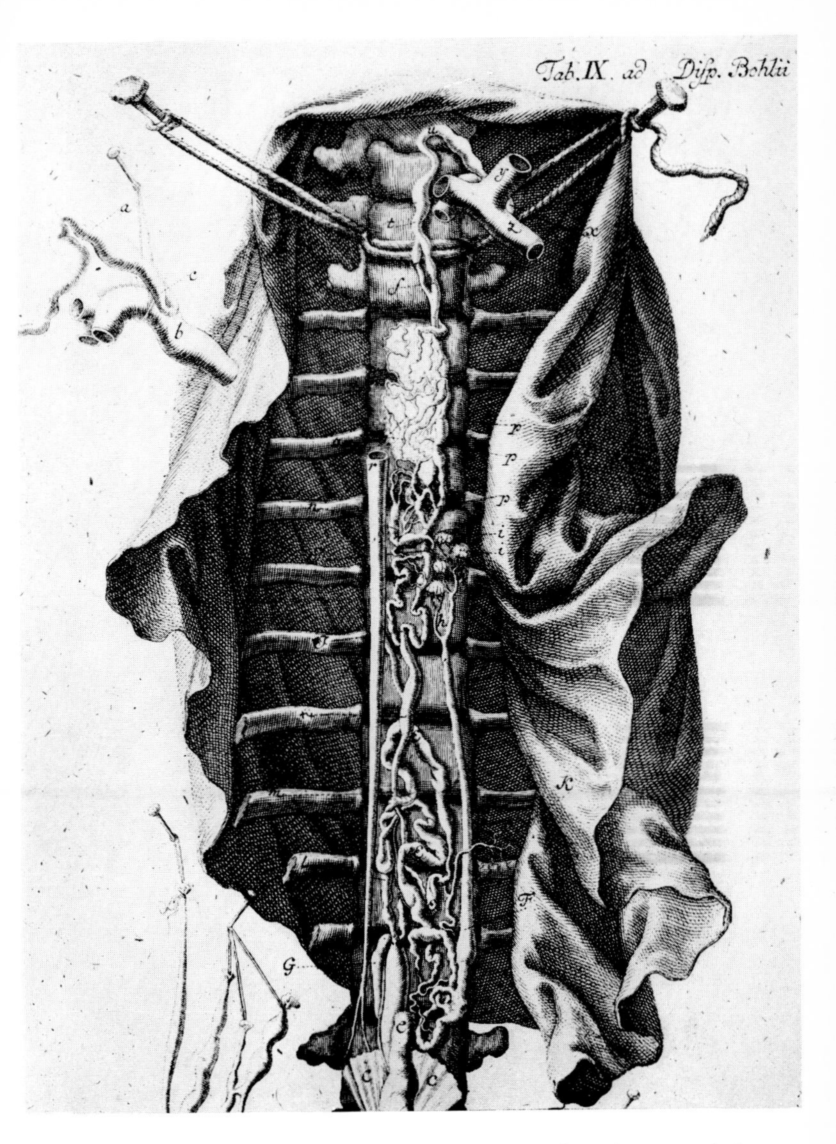

PLATE III A dissection of the human lymph system from the anatomical work of Haller (1746).

See Page 65

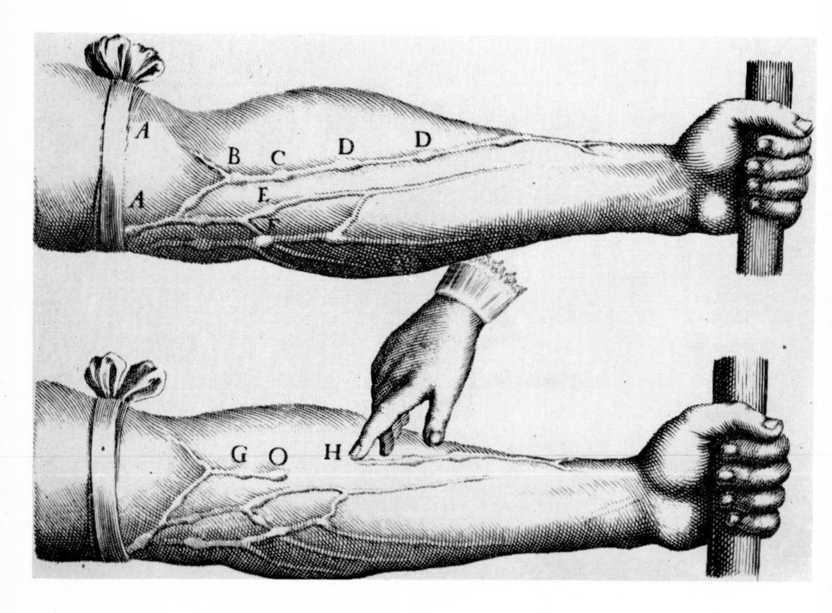

PLATE IV Diagrams from Harvey's work on the circulation of the blood (1628). The flow of the blood in the veins is shown to be towards the heart.

See Page 66

before it was shown that respiration occurs in all living cells, both plant and animal.

The healing of wounds together with extensive investigations in surgery was made by the Scot, John Hunter, at one time consultant at St. George's Hospital in London. Hunter thought that it was in the nature of the body to protect itself from hostile external influences and to repair as far as was possible. He recognized that pus is not a healthy sign in healing and believed that when the nature of this putrefaction came to be fully understood it would be possible to prevent its occurrence (see p. 72).

Hunter's teaching and methods influenced many of his pupils, and the discovery of vaccination by one of them, Edward Jenner of Gloucestershire, is described in the following chapter.

The Origins of Modern Medicine

There are very many branches of science involved in our contemporary practices of medicine, and we shall examine some of the outstanding contributions made in each of the main branches since 1800. Because of its peculiar importance, the biology of micro-organisms and the germ theory of infections disease has been described in a complete chapter.

The major aspects of medical science that will be considered here are the development of anaesthetics, improved diagnostic and surgical techniques, the nutrition of the body and deficiency diseases, endocrines and their disorders, and fundamental progress in physiology.

The Development of Anaesthetics

The use of opium was known to the Greeks, and during the later centuries all sorts of remedies had been used to relieve pain. Alcohol, compression of the part effected, and hypnotism are just a few others, but none were really effective until the introduction of anaesthetics in the eighteenth century. It should be understood that intense pain, such as might occur during an amputation, can bring about a condition of shock quite likely to kill the patient.

It was the advances of chemistry that made anaesthetics available and thus led to more refined surgery. In 1802 Humphry Davy announced the properties of nitrous oxide included its ability to induce lightheadedness and unconsciousness, and some years later Faraday revealed the same properties for ether. These observations were not at once followed by practical application, but it seems that "laughing gas" and ether "frolics" had become quite common by the 1830's. In the following decade a number of trials of both these gases as surgical anaesthetics (Fig. 5.3) were made in the United

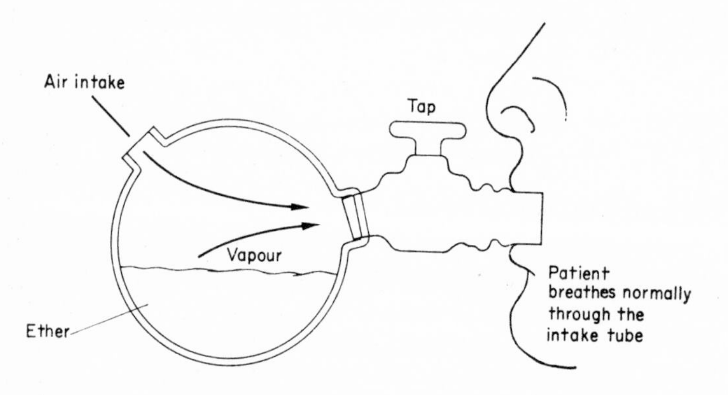

FIG. 5.3. Original type of anaethesizer, *c.* 1850.

States. One of the more unfortunate of these was carried out by Horace Wells of Connecticut, a dentist who was convinced of the advantage of using nitrous oxide in the removal of teeth. When he made a public demonstration of such tooth extraction at Harvard Medical School, the patient was heard to groan during the pulling and Wells was ridiculed. In his private practice he continued to use the gas until one of his patients died under its effects.

Shortly after this, chloroform, a much more powerful anaesthetic than any tried previously, was used by James Young Simpson of Edinburgh. This was followed by Nieman's discovery that injection of cocaine could render local areas of the body insensible to pain.

In our own time general anaesthetics are followed by administration of curare (Indian arrow poison) which prevents local muscular reflexes and allows a much lighter degree of anaesthesia. Besides this, the blood supply to the regions upon which surgery is being done can be cut down by appropriate drugs, and very recently it has been found possible to reduce the metabolic rate of particular organs by cooling. This later technique has proved particularly important in surgery of the brain and heart which have normally a high oxygen requirement which limits the length of time they can be deprived of blood.

While all these advances have been made since Simpson introduced chloroform, it should be remembered that at the time it may have made longer and more complex operations possible but it did not stop the appalling loss of life from post-operative infections. The work of Pasteur and its application by Lister to antiseptic surgery had still to be done, and in the 1840's the chances of recovery from an operation were only some 50 per cent.

Improved Diagnostic and Surgical Techniques

Refinements in the diagnosis of all sorts of diseases were made throughout the nineteenth century. For examination of the eye, Van Helmholtz, also famous for the theory of colour vision, invented the ophthalmoscope. As pathology and micro-biology became widely understood, the microscope proved to be a useful tool in the identification of specific diseases.

Various types of kymograph (Fig. 5.4), an instrument that records

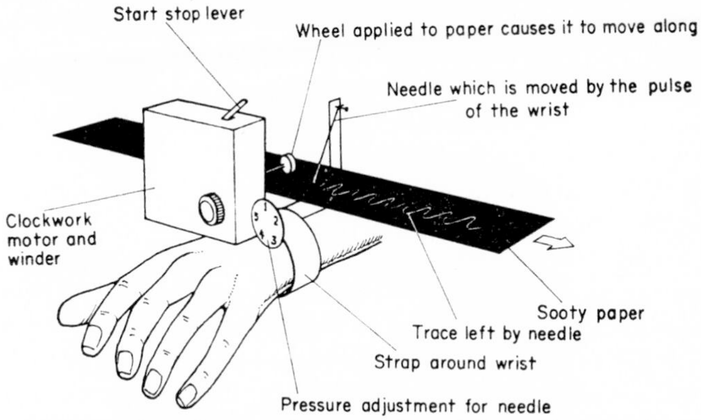

FIG. 5.4. An early type of kymograph used for making records of the pulse.

small movements by tracing on a smoke-covered drum, were used to follow the pulse wave and detect abnormalities.

Outstanding in this field was the application of X-rays, discovered by Röntgen in 1895 to diagnose fractures, tumours, and tuberculous lesions. This marvellous means of looking into the body without having to open it up has become more and more refined until today the introduction of fine tubes through the arteries and into the heart itself can be followed by continuous X-ray photography. Once inside the heart substances such as iodine, opaque to X-rays, can be led in through the tubes and faults, such as holes between the ventricles clearly seen.

Despite the aids to accurate diagnosis, medicine as a whole could never really cope with contagious disease until the true nature of infection was understood. Semmelweiss had suggested that childbed fever, which killed many mothers after the birth of their babies, was an infection spread from one woman to another by the midwifes and doctors ministering to them in their labour. His ideas were to fall on stony ground until in the 1860's Louis Pasteur showed that decomposition was due to contamination by micro-organisms, and that these tiny creatures could actually cause disease. Pasteur's discoveries were put into effect by the surgeon Joseph Lister of Edinburgh, who was already teaching that suppuration of a

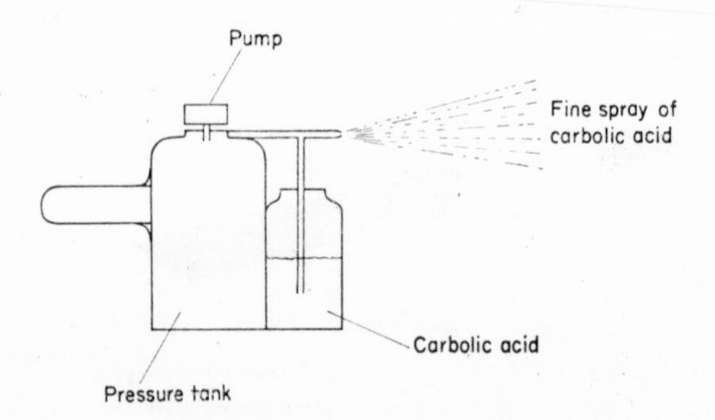

FIG. 5.5. Lister used this sort of spray to cover his operations and also washed his hands and instruments in carbolic acid before starting his surgery.

wound was itself a form of decomposition. It occurred to Lister that if Pasteur was correct about the association of decomposition and microbes substances which destroy, the latter should also prevent wound infections. With this in mind he tried a number of chemicals including carbolic acid, and originated the notion of antiseptic surgery. Lister washed his hands in the carbolic as well as his dressings and instruments and also sprayed the operation site and theatre (Fig. 5.5). This method was remarkably successful in preventing post-operative infections and also did much to prove the validity of Pasteur's theory of the cause of infection. In a letter written to Pasteur he says:

> Permit me to thank you kindly for having shown me the truth of the theory of germs and putrefaction by your brilliant researches, and for having given me the sole principle which has made the antiseptic system a success. If you ever come to Edinburgh it will be a real recompense to you, I believe, to see in our hospital in how large a measure humanity has profited from your work.

Carbolic acid (also called phenol) is a corrosive substance, and although it kills bacteria and in concentration animal cells also, it is more suitable for disinfection of floors or toilets than for application to delicate skin and tissue. The modern method is to use aseptic surgery whereby all germs are excluded from the operating theatre and the persons and instruments of the surgeons are rendered sterile and free from germs.

Despite this attempt to exclude all harmful microbes they do sometimes appear after operations, and the use of antibiotics has also been of great help in post-operative infections.

The ability of the surgeon to deal with the more complicated parts of the body such as lungs, heart, or brain has also increased immeasurably over the last century. In the 1890's repairs were first made to the cardio-vascular system, which includes the heart and arteries. The very substantial difficulties that attend the latter have been at least partly resolved by passing the blood through an artificial oxygenator so that the heart itself can be isolated and worked on.

For surgery on the lungs it has also been found possible to put them out of their normal circuit by collapsing or sidetracking and thus make possible operations that could not otherwise be done. By far the most difficult part of the body for surgery, however, is the brain, and although trephination of the skull goes back to early times, neuro-surgery in its present form dates from the 1890's when Sir Victor Horsley made successful removal of brain tumours. This work was later extended by Harvey Cushing in the United States.

In all sorts of accidents and operations there tends to be a considerable loss of blood, and the transfusion of animal blood into man in the eighteenth century had not been found safe. The reason for this was shown by Landsteiner in 1900 when he discovered that human blood exists in various groups and a mixture of these can lead to the formation of clots which may impede the circulation. The blood of animals also is incompatible to our own and thus cannot be used for transfusion. Once the nature of these blood groups was understood, transfusions between similar groups could be made with safety, and within the last few decades blood transfusion services have been set up and supplies of blood or plasma are available at all times and as needed.

The limitations of surgery are reached when whole organs need to be replaced, for the body will normally accept only grafts from its own tissues and not from any other person except for an identical twin—not even a close relative. (An exception to this is the cornea of the eye which is not provided with a blood supply and can be freely grafted from one person to another. As with blood, a "cornea bank" has been established in the last few years.) There is evidence that the problem of grafting in whole new organs is not insurmountable, and current research is going on aimed at suppression of the reaction, whereby the body rejects foreign protein.*

Nutrition and Deficiency Diseases

The bulk of the solid part of our bodies and of our food consists of the three classes of organic substance known as fats, carbohydrates, and proteins respectively. This fact was demonstrated by the German chemist Liebig, whom we shall be meeting again in connection with agriculture (page 103), and it was also clear that our diet must

*The reader will be familiar with the various attempts, as yet far from successful, in the transplanting of hearts, kidneys, lungs and parts of livers that have been carried out recently.

contain these foods in proper balance. In fact, unknown at the time of Leibig's work, there are two types of protein—the essential which contains constituent amino acids we cannot ourselves make, and non-essential with less valuable amino acids. If one's diet does not contain sufficient of the first-class protein a deficiency disease, such as kwashiorkor (seen in West Africa) may result.

However, it was noticed long before Liebig's time that a disease common on board ships, which showed itself by large spreading haemorrhages and which was known as scurvy, could be prevented if fresh citrous fruit was eaten. Commander Lind brought this practice into use in the British Navy in 1757, and it was also used by the famous explorer Captain Cook on his voyages. (The nickname "limey" applied to us by the Americans originates from this taking of lime juice on board ships of the Royal Navy during the American War of Independence.)

Before the end of the nineteenth century it had been shown that a diet consisting solely of carbohydrate, fat, and protein, that appeared to be ample to sustain life, did not in fact do so. Far from maintaining health, such a diet led to all manner of disorders and eventually to death. At the same time two Japanese workers found that they could actually induce the disease beriberi (previously thought infectious) by feeding chickens on a diet of polished rice. In those parts of the Far East where the disease is found in man a large part of the diet was also polished rice.

It thus appeared that some diseases, grouped generally as "deficiency diseases", could be due to factors lacking in the diet, and with this idea in his mind Gowland Hopkins made a prolonged study of dietary requirements. His work, which was mostly done between 1906–12, showed that although rats could not flourish on a diet of pure carbohydrate, fat, and protein, they could be reared successfully if very small quantities of milk were added to their food each day. Hopkins suggested that milk contained substances, present in small amounts, that were essential factors in nutrition, and in 1911 Casmir Funk isolated such factors and called them "vital amines". The factors are not all amines as Funk supposed, and the name was subsequently changed to its present form of vitamin.

At first these vitamins were called after the diseases they prevented, so there was an anti-scurvy vitamin, an anti-ricketts vitamin, and an anti-beriberi vitamin. During the First World War the alphabetical system of naming vitamins was introduced and since that time we have not only the original A, B, C, and D but many subdivisions as well as vitamins E, F, K, P, etc.

Besides the essential proteins and vitamins, a healthy diet must also contain certain mineral salts, and in this century the prevalence of mineral deficiency diseases, such as simple goitre, in regions where iodine is lacking, have been successfully combated.

Endocrines and their Disorders

The chemists and physiologists had shown the body to be a complex chemical machine whose various functions could be studied in isolation, but at the same time it was clear that these processes within the living body were closely integrated and regulated.

One sort of regulation was by means of the nerves, and in the early nineteenth century the work of Magendie and Bell had shown how sense organs and effector organs were "wired" up by means of nerves passing across and along the spinal cord. The basic unit of nervous integration, that is the reflex arc, was described by these workers, but the concept was much extended and the nature of autonomic control discovered by the researches of Sherrington in this country and Pavlov in Russia. Pavlov was particularly interested in the modification of reflexes by conditioning, and his work on the salivary responses of dogs is well known.

Besides co-ordination of the body by nerves, another system was also found to exist, particularly associated with long-term changes. This was the system of ductless glands or endocrines, whose secretions—the hormones—passed directly into the blood stream and thence were carried to all parts of the body.

Thomas Addison (1793–1860) had described a set of symptoms which included muscular weakness and paralysis which he associated with a malfunctioning of the adrenal glands, a pair of small organs found one above each kidney. These glands, together with the pituitary at the base of the brain and the thyroid in the neck, had been noticed by earlier writers but their functions were unknown.

Following Addison other workers described another set of symptoms, namely fatness, sluggishness, and retardation which were produced by the malfunction of the thyroid. The opposite symptoms were also associated by Graves with over activity of this gland. The former disease, together with its congenital form, called cretinism, could be cured if regular quantities of thyroid gland were eaten, but it was not understood why this should be so.

The first hormone to be actually identified was a rather obscure one which is produced in the walls of the duodenum and travels to the pancreas, stimulating release of its digestive juices. This hormone, discovered by Bayliss and Starling in 1902, was called secretin.

A very widespread endocrine disease is diabetes, where sugar regulation does not take place and coma and death can result. The detailed structure of the pancreas had been revealed by Langerhans in 1869, and some correlation between this organ and diabetes was made by the finding that removal of the pancreas in dogs led to the onset of this disease.

It was not until 1921 that the hormone from the pancreas was isolated and then, 6 years later, it became available for the treatment of diabetics. The men who did this were two Canadians Banting and his assistant Best. They knew that people who died of diabetes showed degeneration of the pancreas but that so far it had not been possible to extract a hormone from the gland. They suggested that a tying of the pancreatic duct, whereby its digestive cells destroyed themselves, would leave the hormone intact. This proved to be the case, and they extracted the hormone and called it 'isletin" because it came from the islet cells of the gland. Later the name was changed to insulin and, prepared from cattle, it is now available to diabetics and allows them to lead a normal life. Insulin is also of interest in being the first large protein whose molecular structure was determined (Sanger, 1954).

General Advances in Physiology

Claude Bernard (1813–78), the great French physiologist, said: "Medicine is the science of sickness, physiology is the science of life;

thus physiology must be the scientific basis of medicine." Certainly since the beginning of the nineteenth century, fundamental discoveries in physiology have preceded or gone hand in hand with developments in medical techniques.

Bernard himself saw that the body maintains itself in a constant state in respect of temperature, acidity, hydration, salts, oxygen, and wastes. He made some attempts, since much amplified to explain the nature of these various controls.

In almost every field of physiological investigation very large advances have been made since the early nineteenth century.

The relationship of the autonomic system which works such organs as the stomach and heart to general activity of the nervous system was shown by Dale and others in the 1920's.

All living cells require constant supplies of energy and how this is obtained from carbohydrate and other substrates has been revealed in the decades since the last war. Sugars were found by Krebs to go through complex series of reactions and to give up their energy, a little at a time to organic phosphate compounds. These latter, such as the now well-known adenosine tri-phosphate, yield immediate energy as required.

In 1963 Hodgkin and Huxley were awarded Nobel prizes for a long series of researches which explained how the nerve impulse was generated and travelled. Within the brain itself the functions of large regions have been mapped out. About 60 years ago almost nothing was known of the functioning of this all-important organ. Needless to say, medical ignorance is still greatest in this field.

Movement is one of the characteristic physiological activities of living things, and the nature of striated muscle and its method of contraction have also been revealed during the past decade.

With new techniques taken from physics and chemistry, the molecular structure of large natural proteins such as myoglobin and haemoglobin has been elucidated. Work on the code of the genes (see p. 57) has brought us near an understanding of the whole functioning of living protoplasm, and as this understanding grows we may perhaps expect the problem of cancer—abnormal functioning of the cell—to be solved.

Summary

More than any other branch of science medicine has contributed to the welfare and happiness of mankind. The expectation of life in this country has risen steadily since the eighteenth century, and the infant mortality declined. As short a time ago as 1890 some 170 babies died out of every 1000 born alive before reaching the age of 1 year; nowadays only 20–30 die in the same time in this country. The incidence of all infectious diseases has greatly declined (see Fig. 5.6) and some such as diptheria have been rendered virtually extinct in the human race.

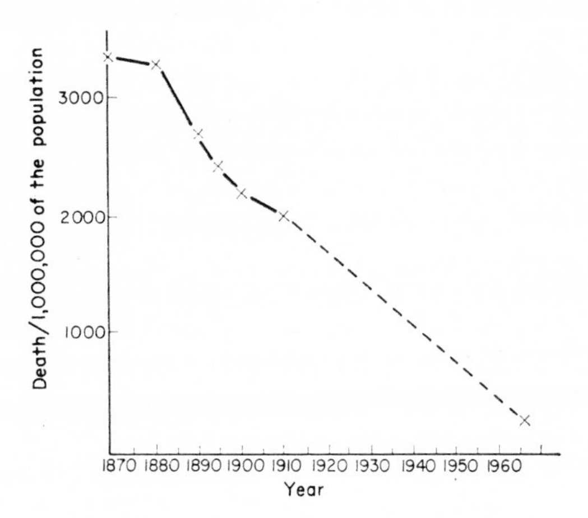

FIG. 5.6. Decline of tuberculosis in Great Britain since 1870.

One can afford vague beliefs and partial understanding of all sorts of subjects and it will do little harm. It does not, in fact, matter so very much to an individual whether the sun goes round the earth or the earth the sun. But this is not true of medicine; practice must be founded on a correct understanding of the nature of disease if it is to be effective.

The application of science to the maintenance of human health has been one of its major contributions.

Questions on Chapter 5

1. What were the main contributions of Galen to medical science and why was it that his views, though often erroneous, persisted for so long?
2. Discuss the importance of Harvey's work on the circulation of the blood.
3. Write briefly on the work of Lister, Banting and Best, and Simpson.
4. What are deficiency diseases and how did they come to be described and treated?

CHAPTER 6

Aspects of Medicine

Part II: *The Discovery of Germs and the Fight Against Infectious Diseases*

THE study of micro-biology embraces those organisms such as the protozoa, bacteria, viruses, and certain fungi whose activity, owing to their small size, can only be followed by use of a light or electron microscope.

The importance of these minute creatures is out of all proportion to their size as they include many disease causing species which plague man, his domestic animals, and his crops. Study of the micro-biology of food and milk is also important for their safe handling and preservation and of the minute organisms of the soil to determine conditions for maximum fertility. Surveys of water bacteria and proper disposal of faeces prevent contamination of our drinking supplies.

Recently the secretions of certain fungi have been used as antibiotics whose generation and activity are important to medical science while, at the molecular level, modern research into the structure of micro-organisms is providing us with new knowledge about the organization of living matter.

The First Discovery of Microscopic Organisms

Although early philosophers had suggested sub-visible "germs" as existing and even as causing disease, such entities remained theoretical until the discovery of the microscope.

In the thirteenth century Roger Bacon published a book on lenses and the science of optics, and by the seventeenth century the grinding of lenses was being practised in various parts of Europe including Holland. Janssen (*c.* 1600) made the first compound microscope, but the instrument was too inefficient to see micro-organisms. It was left to Antony van Leeuwenhoek, another Dutchman, to reveal the world of the microscope in 1676.

Van Leeuwenhoek was a man of little education and was employed as a janitor in his native town of Delft. He ground his own glass and made simple microscopes (Fig. 6.1) employing only a

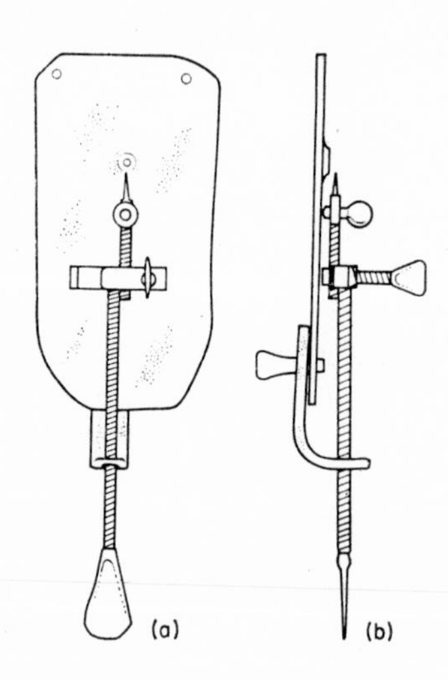

FIG. 6.1. Leeuwenhoek's microscope.

single lens. Despite this simple arrangement, some of his instruments must have had a magnifying power of nearly 300 diameters to have enabled him to see the objects he described. For a period of 47 years van Leeuwenhoek communicated his findings to the Royal Society in London in a series of letters. Among the many things he described were the main types of bacteria, many Protozoa, human sperm and blood corpuscles, and the capillaries that join up arteries and veins. He also recognized the cellular nature of wood (although Robert Hooke, another early microscopist, described the cellular nature of plant material some time later).

Despite the accuracy and extent of his observations, van Leeuwenhoek's discoveries relied on his own private techniques, and work of a comparable standard was not done for nearly another century.

Gradually the resolving power of the microscope was improved, and the names of Robert Hooke, (Fig. 6.2), Harris, Joblot, and John

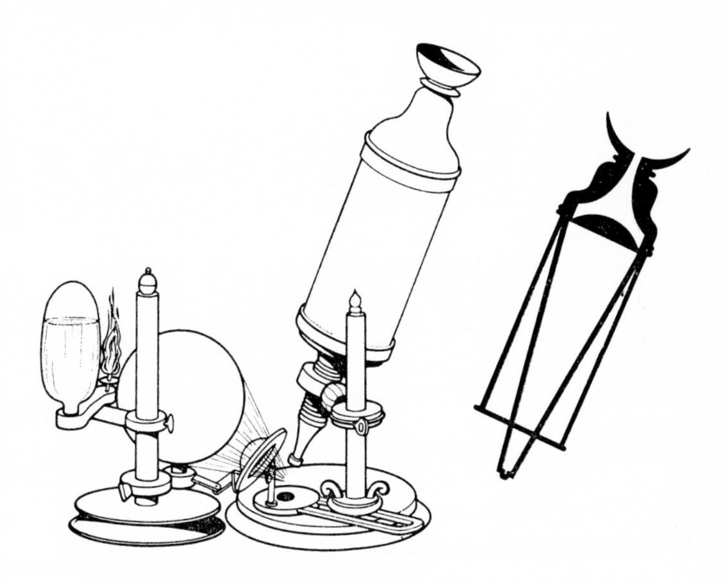

FIG. 6.2. Robert Hooke's microscope was more complex than Leeuwenhoek's, but suffered from chromatic aberration. There had to be intense source of light focused on his material.

Hill are associated with microbiology in the period following the death of Leeuwenhoek. These men classified and described a number of protozoans including *Euglena*, *Paramecium*, and the dinoflagellates. In his great classification, the *Systema Natura*, published in 1735, Linnaeus attempted to delegate protozoans to the classes Volvox, Furia, and Chaos. Only one of these groups is still recognized.

By the beginning of the nineteenth century the microscope was much improved and the major problem of chromatic aberration at high magnification was solved. By 1838 sufficient information on microscopic organisms was available for Ehrenberg, of Leipzig, to publish a comprehensive work on the protozoa in which many hundreds of species were described with a high standard of accuracy.

The Controversy over the Spontaneous Generation of Life

With the discovery of the world of microscopic life the age-old controversy about the spontaneous generation of living organisms from inanimate matter could be resolved. In the mid-sixteenth century Francesco Redi had shown convincingly that maggots cannot appear in meat without the visitation of female flies, and the whole idea of the generation of higher forms of life spontaneously was not seriously regarded after this time. Otherwise, however, was the generation of micro-organisms which was still thought to occur spontaneously in favourable conditions.

This problem was first investigated by the Italian priest Spallanzani who in the year 1777 broiled a broth in glass containers and sealed these from the air. He was able to show that in such conditions no life appeared in the broth, but his researches were not entirely acceptable to his contemporaries. Particularly strong opposition occurred in England where another priest, J. T. Needham, suggested that Spallanzani's experiments had destroyed a "vegetative force" which was necessary for the animation of the broth. Despite this criticism, Spallanzani's method of sterilization found application in Appert's method of preserving food by canning which was introduced in 1810, and won a prize from Napoleon for the application of science to human wellbeing.

Further investigations into the spontaneous germination of life were done by Schwann who in 1837 passed air over red-hot tubes into boiled soup infusions and showed that, although oxygen was present, no micro-organisms developed. In fact it was left to Louis Pasteur to destroy, once and for all, the notion that life could originate spontaneously and this he did in 1864. Pasteur's experiments were simple but ingenious. He boiled up soups in flasks with long goosenecks (Fig. 6.3) and although the flasks were not sealed their contents did not putrify; microbes which fell into them accumulated in the bend of the neck. On shaking, so that some of the contents spilled into the neck, putrefaction set in. Not only did Pasteur discredit spontaneous germination but, by opening a number of sealed flasks in a whole set of environments, from a Paris cellar to the slopes of Mount Blanc, he showed that microbes, or their spores, were present throughout the earth's atmosphere.

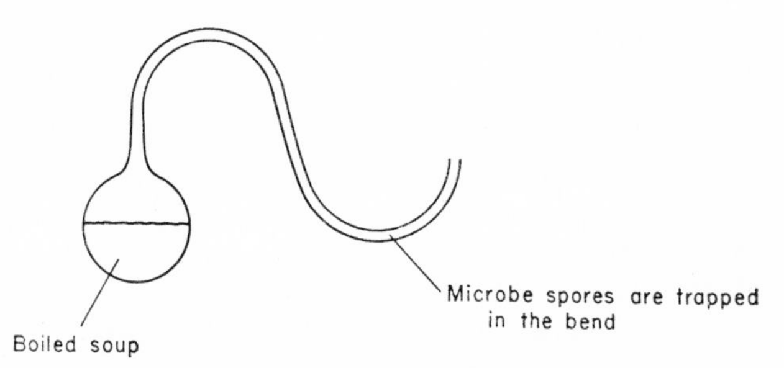

FIG. 6.3. Pasteur's flask.

The Connection between Micro-organisms and Disease

As more microbes came to be discovered and some of these from human intestines and skin, the obvious suggestion that these organisms might cause disease was put forward. This idea was attributed to von Plenciz in 1763 although it is worth recording that it had originally been suggested as early as 1546 by Fracastorius.

The first implied use of the notion that disease could be transferred by infectious material was made by Jenner who in 1796 vaccinated individuals against smallpox using the very much less virulent cowpox. Despite the effectiveness of the treatment it was based on Jenner's general observation that those who had suffered from cowpox were rendered immune from smallpox and the causative agent of these diseases was not seen at this time. In fact a practice of immunization against smallpox by exposing oneself to the disease had been used in certain Mediterranean countries such as Turkey as well as earlier in China.

A direct connection between an observable micro-organism and an actual disease was made by Johann Schönlein who succeeded in isolating the parasitic fungus which causes favus (a kind of scalp ring-worm) in man. This work was done in 1839 and was followed 14 years later by De Bary's investigation into the moulds, smuts, and rusts of crop plants (see p. 113).

Working in the wine distilleries of Lille the young chemist Louis Pasteur found the bacterial infection that was killing the yeast plants and spoiling the wine. He also described the bacteria that can turn wine to vinegar under certain conditions. The methods he used of pure culture of yeast and "pasteurization" of the fermented liquor by gentle heating are still used today. From the diseases of plants Pasteur proceeded to those of animals and was able to isolate and recommend a means of combating the disease (pébrine) of silk-worms that was sweeping through southern France. By 1865 Pasteur was convinced that the connection between micro-organisms, especially bacteria, and disease must extend to the higher animals and man. Indeed, such a conviction was growing elsewhere as in the precautions laid down by Semmelweis for the prevention of puepural fever after childbirth (see Chapter 5). A direct application of Pasteur's ideas was Lister's development of antiseptic surgery about which more will be found in the previous chapter.

Credit for the isolation of the first causative agents of a human bacterial disease goes to the great German bacteriologist Robert Koch. In 1876 he isolated the bacteria which causes anthrax. Koch's success as a microbe hunter was due to the culture methods, such as

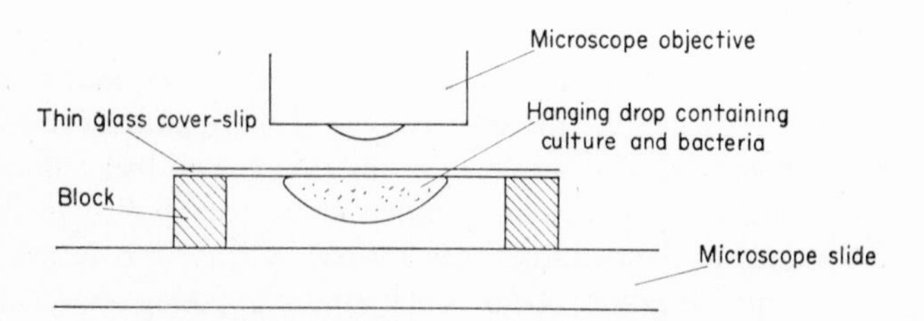

FIG. 6.4. The hanging drop technique used by Koch for culture and observation of bacteria.

the hanging drop (Fig. 6.4) and the nutrient jelly plate, that he invented. By such methods he was able to isolate single species of bacteria and study their activities as pathogens. The bacteria responsible for tuberculosis was isolated, again by Koch, in 1882 and cholera by one of his pupils a few years later. Throughout this period the methods of staining bacteria as a means of identifying and classifying them were being developed.

The next major step in the history of microbiology and disease was Pasteur's perfection of a technique for weakening harmful bacteria and using them to provide immunization against a specific disease. The first experiments were made with chicken cholera where he found that old cultures of the microbes became so weakened or attenuated that they did not cause the disease when injected into a new host. These old cultures did, however, render such a host immune from the disease. Similar procedures were next developed using anthrax bacteria and eventually an immunizing vaccine prepared. The efficiency of the vaccine was demonstrated by Pasteur in a great public field trial using a number of farm animals including sheep and horses. The control animals had no immunization, and on being given injections of anthrax they contracted the disease and died. A similar number of animals were injected by Pasteur with gradually increasing doses of his weakened strain and after three such innoculations they were given the full strength virulent organisms. These latter animals showed no sign of the disease. After this the anthrax vaccine was used all over the Continent and virtually put an end to this widespread disease.

It remained only to apply these methods to a human disease, and the opportunity came from Pasteur's investigations into rabies. Although this disease, which is spread by the bite of infected dogs, is caused by a virus and hence was invisible to these bacteriologists, it was found possible to weaken it as had been done with cholera and anthrax. The method used was to inject the saliva from a mad dog into a rabbit which developed the disease. The spinal cord from the dead rabbit was dried and an extract of this introduced or put into another animal. After this procedure had been repeated a number of times the virus was weakened (in fact a less virulent strain had developed) so much that it no longer caused the disease although it caused the animal to become immune. In 1885 a boy, Joseph Meister, who had been bitten by a mad dog, was brought to Pasteur and was given a series of inoculations of the weakened rabies preparation.

The boy did not develop the disease and a new frontier had been crossed in mankind's war against the microbes.

Five years later von Berhing and Roux, the latter a colleague of Pasteur, developed an inoculation against diphtheria and a further one against tetanus.

The Basis of Immunity and the Beginnings of Chemotherapy

Following the brilliant discoveries of Pasteur and Koch an interest began to develop into the means whereby the body resists disease and how it may be helped artificially.

By a chance observation the Russian biologist Elie Metchnikoff observed the activity of phagocytic cells in the larvae of starfish. He saw that wandering amoeboid cells within their bodies ingested food and other particles and attacked splinters of wood that he introduced into the organism. He found similar cells in the blood and invented the term phagocyte to describe them. Metchnikoff believed that the method of phagocytosis was the only means of destroying microbes in the body while other workers, such as the German Ehrlich, believed that the body produced chemicals against invading germs. We know now that both these methods operate, but at the time there was much controversy on the matter.

This same Paul Ehrlich began a long series of experiments into the chemotherapeutic (literally healing chemicals) properties of various organic compounds including aniline dyes. Already Laveran, who discovered the malaria parasite, had tried injections of arsenic into mice infected with trypanosomes and found that the latter were killed. Unfortunately most of the mice died from arsenic poisoning. Finally, in 1909, a less toxic arsenical compound was tested, namely dioxydiaminoarsenobenzoldihydrochloride, or as it is more commonly known 606 or Salvarsan. This was found to be effective as a cure for syphilis. Another compound, Trypan 3, was developed about the same time against trypanosomes (see p. 92).

Progress in chemotherapy was slow in the succeeding years and it was not until the mid 1930's that Domagk presented the world with sulphonamide, a modified azo-dye active against a number of bacterial infections including pneumonia, meningitis, and gonorrhoea. The discovery of this compound revived interest in chemotherapy and in 1939 the firm of May & Baker sent a chemical, number 693, for testing. This substance was subsequently known as M & B and was found to be effective in treating pneumonia. Strangely enough its bactericidal properties were worked on by Alexander Fleming whose name was to be associated with the much more effective antibiotic penicillin.

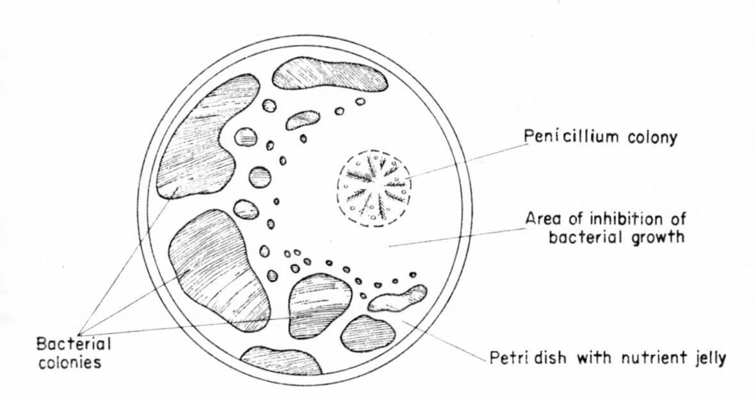

FIG. 6.5. Chance observation of the effects of penicillium on bacterial growth led to the discovery of antibiotics.

The story of penicillin is a remarkable one, as its properties were discovered some 10 years before it was actually developed as an antibiotic. In the late twenties Alexander Fleming noticed that the mould, which had accidentally infected one of his culture plates, caused a suppression in the growth of the surround micro-organisms (Fig. 6.5). Although Fleming had done some preliminary work, it seems that little notice was taken of the observation at this time or indeed until it set off the researches of Florey and Chain into penicillin in 1939. These two workers rapidly demonstrated the effectiveness of penicillin against a wide range of bacteria, particularly the streptococci and staphylococci which cause many diseases in man. They also found how to extract it in large quantities for commercial use.

Further investigation into a wide range of soil fungi yielded the powerful antibiotics actinomycin (1940) and streptomycin (1942), the first chemical that would destroy the bacteria that cause tuberculosis.

The Virus as an Agent of Disease

The difficulty of isolating specific causative agents for a number of important diseases such as yellow fever, smallpox, and polio was resolved in 1892 by Iwanowski. He demonstrated that the mosaic disease of tobacco was caused by a sub-microscopic filterable agent which he termed a virus. This work naturally opened men's minds to the fact that a variety of other diseases could be due to virus infections. Besides being ultra-microscopic at least to light microscopes, viruses only grow on living tissues so cannot be investigated in the same way as bacteria.

Six years later Loffler and Frosch showed that hoof-and-mouth disease of cattle was caused by a virus, and during the first two decades of the present century a number of other diseases had been tracked down to these organisms. Bacteriophages, the viruses that attack bacteria, were found in 1917, and by 1931 a method of growing viruses on living media, such as hen's eggs, had been developed by Goodpasture. Once the viruses could be grown it was only a matter of time before vaccines were prepared. One of the most famous of these is the Salk vaccine against poliomyelitis presented in 1957.

The viruses were first seen with the electron microscope which came into general use in the mid 1940's (Fig. 6.6). Recent years have seen the development of further vaccines, and at present the properties of a substance called "interferon", prepared from living cells which counters viral invasion of the cell, are under investigation.

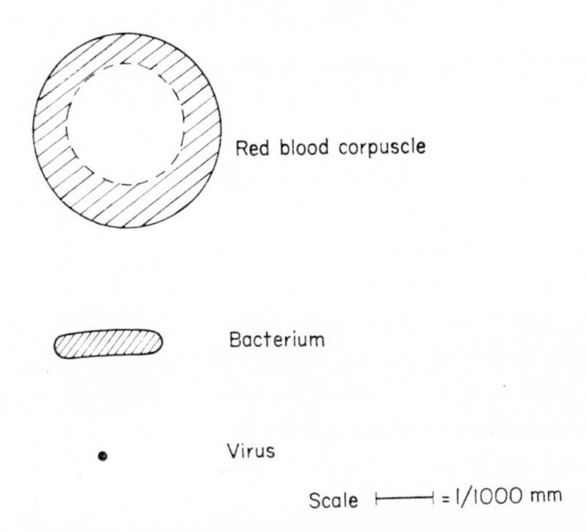

FIG. 6.6. The relative sizes of bacteria, viruses, and cells. Viruses are so small that they cannot be seen with ordinary light microscopes.

Insect and Other Vectors of Diseases

It was not only important to identify specific disease-causing organism but also the means by which they were spread. One important method of disease transmission turned out to be by animal vectors. A vector is an animal by which a disease is transmitted from one host to another, and of the many that exist the malarial mosquito, the yellow fever mosquito, the rat flea, the tsetse fly, and the louse are of most importance to man. A great factor in the elimination of vector-borne diseases is in the destruction of the vector itself.

The first vector of disease can be said to owe its discovery to Patrick Manson whose name was later to be associated with malaria. In 1879, while working in Hong Kong as a medical practitioner, Manson found the young nematode worms that cause the disease filariasis inside a mosquito. This discovery was to influence his thinking when he returned to England and cause him to suggest to the Indian Army Doctor Ronald Ross that he should search for the malarial parasite in the mosquito. This parasite, which is a protozoan, was seen in human blood by Laveran in 1880.

Meanwhile the work of Theobald Smith on the Texas fever of cattle had shown by 1893 that the disease is passed from one animal to another by the bite of a tick. A year later Yersin and Kitasato, working in the plague epidemic that was raging in Hong Kong, identified the plague bacillus and showed it to be the same as that of rat plague. Some years later, by which time the bubonic plague had spread to India, the plague commission of that country tracked down the vector of the germ to the rat flea.

In Africa, again in 1894, David Bruce was the first man to see trypanosomes in the blood of animals and 6 years later to find the species of *Trypanosoma* which caused sleeping sickness in man (Fig. 6.7). By 1903 he had identified the tsetse fly as the vector of this disease whose presence had made great parts of Africa almost uninhabitable by man.

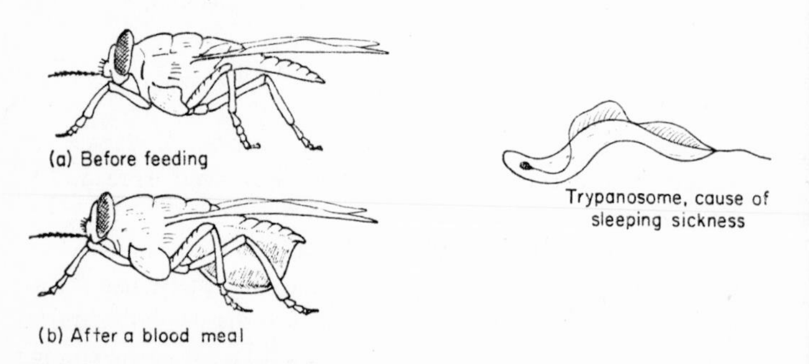

Fig. 6.7. The tsetse fly and the trypanosome that it carries.

These last two decades of the nineteenth century were an exciting time in the revealing of vectors, and of the many discoveries made the finding of the malarial mosquito is probably best known. As stated above, the credit for this is due partly to Manson and partly to Ross as well as to the Italian Grassi. Acting on Manson's advice, Ross examined a large number of mosquitoes for traces of the malarial parasite, finally hitting on a species that carried bird malaria in

1895. It took him a further 3 years to finally prove that the mosquito actually could transmit the disease, and his findings were paralleled and reinforced by those of Grassi working in the malarious swamps outside Rome. The mosquito was now receiving attention in other parts of the world, and in the year following Ross's and Grassi's findings, Walter Reed, working with an army research team in Panama, proved the species *Aëdes aegypti* to carry the virus yellow fever.

Since those first vectors were discovered the list has grown and includes the lice which carry typhus (1916) and the sand flies and mosquitoes which carry various worms and skin diseases. It also includes aphid vectors of many plant diseases.

Since the Second World War health organizations (particularly WHO) have tackled the elimination of vectors in many parts of the world—often with great success.

Although modern insecticides are highly effective in the destruction of insect pests, the latter show surprising speed in the breeding of resistant strains, and malaria and worm infestation, spread by insects, make up the most important diseases in the world. In other roles insects eat a great deal of human food in a world where many are starving.

It is obvious that a great deal remains to be achieved.

Questions on Chapter 6

1. What contributions did Leeuwenhoek make to micro-biology?
2. How did the notion that life could originate spontaneously come to be disproved?
3. What were the contributions of Pasteur and Koch to the fight against infectious diseases?
4. Write shortly on antibiotics, viruses, and vectors.

CHAPTER 7

The Development of Agriculture

THE techniques of tilling the soil, sowing, hoeing, and harvesting, as well as the rearing and care of animals, are as old as the history of civilization itself—indeed, without them there would be no civilization.

A comparison of the records of Egyptian farming from 6000 years ago with those of the early sixteenth century show that there had been relatively little change or advance in agricultural practices and principles certainly in comparison with the spectacular progress made in the contemporary agricultural revolution. This latter began with the effective application of science to agriculture in the eighteenth century. For all the ingenuity and analysis of great Roman landowners and writers on agriculture, and for all the activity of the medieval ecclesiastical and secular estates and even the development of a monetary economy, the fact is that agriculture before the sixteenth century was largely an exercise in self-sufficiency. In those times an average of 4 cwt of cereals to the acre as the yearly yield required the ceaseless toil of the majority of humanity. Now we expect yields of between 25–60 cwt an acre in western Europe and other parts of the world where scientific methods are applied. This difference in return has meant the increase, the liberation from the soil, and the enrichment of the human species as a whole.

This brief survey of agriculture will try to follow the ways in which science has been applied to farming and the changes in method from early times to the present day. It is now possible by better seeds, by mechanization of sowing and harvesting, by chemical weed-killing, and manuring for one man to do the work that once required a whole village and to do it more efficiently.

Many of the applications of science to agriculture were first done in this country, and a history of British agriculture parallels much of its changing pattern and efficiency throughout the world.

The Beginnings of Agriculture

Stone Age man was a hunter, and lived for the most part on the meat of wild animals, implementing this diet with roots and other vegetables. In this hunting the comparatively puny strength of man was made up for by his ever-increasing intelligence, his use of stones and spears, and by co-operation with his fellows, but life must have been hazardous and brutish, and the human species very small in numbers.

From this stage and after many hundred thousands of years we find the appearance of nomadic shepherd peoples such as the Israelites whose history is recorded in the Old Testament. The great flocks of domestic animals that belonged to the patriarchs wandered from pasture to pasture, and their meat, milk, and skins provided their owners with food and clothing and tents. It is recorded that Job, and later Isaac, besides keeping animals, practised husbandry or cultivation of the soil, but this latter method of obtaining food was by no means as extensively developed at this time as among the neighbouring Egyptians.

Around the deltas of great rivers such as the Nile and the Euphrates, grew up prosperous civilizations who practised the deliberate cultivation of vegetable crops, mainly wheat and barley, as well as the rearing of animals. These people were not nomadic but remained farming the same pieces of ground for centuries and millennia. It seems unlikely that the need to return nutrients to the soil was understood and the reason for its continued fertility in these regions was the periodic flooding of the great rivers and the deposition of new layers of rich alluvium.

One of the main problems confronting the Egyptian farmer was how to maintain a supply of water to his land. Complex irrigation channels were made and methods devised for raising water from river level to more distant fields. In essence many of these methods, which used buckets on lever arms or water-wheels, are seen in modern Egypt. So good was the soil and so favourable the climate that two crops could be raised each year, and wheat, barley, and millet were grown. As well as plants, cattle, sheep, goats, and fowls were raised.

The plough used by the Egyptians was made of wood and was little more than a sort of hoe which broke up the surface of the soil but did not turn a furrow. It was drawn by one or two oxen. Seeds were scattered broadcast and trodden in by the feet of sheep which were allowed to roam over the land at sowing time. At harvest time a simple form of sickle fashioned from the jaw bone of an animal or of wood set with flints was used and the crops cut high up the stalk. After drying in sheaves the plants were trampled out on hardened threshing floors by the feet of oxen and later the seed separated from the chaff by winnowing.

Cereal crops were then stored until actually required for milling which was done either between two stones or by a more efficient rotary arrangement. The mill could be worked by hand or by animals, and in its more elaborate form consisted of an upper stone with a conical cavity which fitted over a lower cone-shaped stone. Corn was fed in through a hole in the centre of the upper stone, and by the time it had worked its way out through the revolving stones it was ground down to flour.

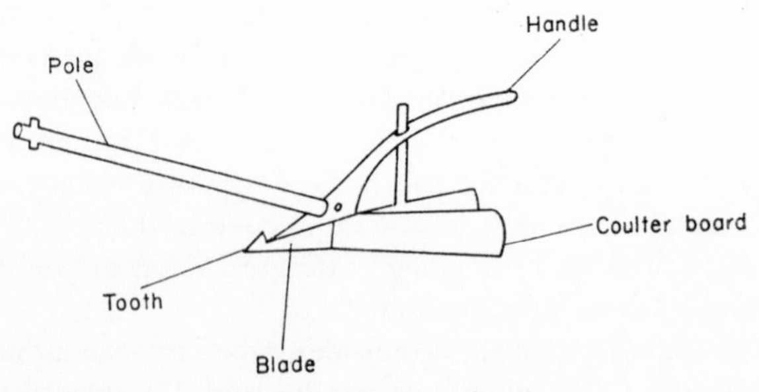

Fig. 7.1. A Roman plough.

With the Romans there was a marked improvement on previous practices. Writers like Pliny and Varro reveal a concern for utility and an interest in increasing yields and lowering costs. Though two-course rotation was normally used, the Romans knew of three-course rotation and appreciated the importance of growing leguminous crops. In favourable areas, such as northern Italy they were able to keep land under continuous cultivation. The light wheel-less plough (Fig. 7.1) was the main instrument of cultivation, but this was to be at least partially replaced by the larger iron-shod plough with wheels and a coulter board which turned a furrow. These larger ploughs were drawn by a team of up to eight oxen, and it was customary to make furrows some 220 yards long and approximately twenty of these could be ploughed in a day. The land covered by such a day's work was called a yoke, and is more or less equivalent to the British acre.

The Romans clearly understood the importance of repeated working of the soil and made use of the harrow as well as lime to improve its fertility. In the cultivation of fruit trees and vines they were so expert that their methods have lasted to our own times.

Farming Methods in Britain from Saxon Times to the Seventeenth Century

The Roman method of agriculture which prevailed in Britain and other parts of their Empire were replaced in this country with the arrival of the Saxons. These people brought the three-field rotation system and common pasturage that were to dominate English farming until the various enclosures of the land that took place between the twelfth and eighteenth centuries.

In essence the system involved a village farming three very large fields, these being divided up into many strips which were allocated to individuals. Thus each man would own a number of scattered strips in each of the village fields and all would have access to the areas of open uncultivated land where livestock could be pastured. In all the holdings of the individual would amount to between 15–30 acres.

On the great fields the crops were rotated on a yearly basis so that autumn-sown wheat was followed by spring corn, barley, oats or peas, and beans, and this by a year of lying fallow. During the period between sowing and harvesting the fields were fenced in, but after the latter the fences were removed and livestock allowed to graze on the stubble. Elaborate laws were made to ensure control over cattle, and manorial courts set up to levy fines on those who contravened these laws. Besides cattle, pigs and poultry were kept, but at this time there were no proper winter feeds for those livestock who could not fend for themselves, and most of the animals were slaughtered in the autumn.

Although this system worked for many hundreds of years it was not really very efficient because of the disjointed allocation of land and the common grazing allowed after harvest. All had to conform to the same pattern, and there was no room for individual enterprise or improvement of methods. The peasants, or villeins as they were called, were bound to contribute their labour to the lord of the manor from whom they derived their rights to the land.

From the twelfth century onwards the stable structure of this medieval society began itself to break up under the pressures of social and economic changes. Money became more plentiful and wages began to replace bonded service in the thirteenth and fourteenth centuries. The Black Death in 1348 killed nearly one half of the population of England, and the availability of labour consequently declined. Because of this and because of increasing wages, many landowners began to fence in their land and keep sheep which yielded a good profit for less work. (This enclosure of individual fields and estates must be distinguished from the later enclosure of common land.) Between 1194 and 1666 the English wool trade flourished to such an extent that the whole prosperity of the country depended upon it. This great age of sheep farming is still commemorated by the fact that the Chancellor of the Exchequer sits on a woolsack in the House of Commons. At the height of this period some 150,000 sheep are recorded from East Anglia alone with a fleece averaging 3 lb in weight and fetching between £2–£9 a sack (364 lb) depending on its quality.

During all these centuries there had only been very gradual changes in farming techniques. By 1650 the practice of autumn ploughing was brought in: carrots, turnips, and other roots as well as cabbages were grown, and clover was used to reclaim barren land. Horses had replaced oxen for drawing the plough, and more ambitious drainage schemes were employed. In a book on husbandry published at this time we find the following advice:

> In thy tillage are these special opportunities to improve it, either by liming, marling, sanding, earthing mudding, chalking, etc.—indeed, everything that has any liquidness, foulness, saltness or good moisture in it is very natural enrichment to almost any sort of land.

Here is a small beginning of the scientific thinking that was to cause such a great revolution in agriculture between the end of the Middle Ages and our own times.

Progress in Agriculture up to the End of the Eighteenth Century

An outstanding contribution to farming practice was the invention of the drill by Jethro Tull (1701). This instrument sowed seeds in rows and at a given depth which allowed much more easy weeding and greatly increased yields. Tull believed that plant roots took in minute grains of soil, and to this end stressed the importance of making a fine tilth. His seed drill incorporated the principles of an adjustable harrow and was pulled along by a horse (Fig. 7.2). The Suffolk drill, very much like Tull's original invention, is still used on many farms today.

Before the drill, sowing had been by hand or broadcast and this method was exceedingly wasteful. A contemporary verse had it "one for the pigeon, one for the crow, one to wither, one to grow" and the randomly positioned plants difficult or impossible to weed. The drill was an agricultural innovation of the very first importance.

A solution to the problem of winter feeding of livestock which had so vexed the medieval farmers was suggested by Sir Richard Weston in 1660 who cultivated turnips for the purpose. Despite the soundness of this scheme it was not generally employed until the more widely known work of Lord Townshend or "turnip Townshend" as he became called. He introduced the Norfolk four-course rotation whereby a root crop was grown between cereals and a fallow year omitted. Townshend made use of the newly invented drill to sow his crops, and he and another Norfolk farmer, Thomas

Coke, were both instrumental in spreading the use of new machines and methods and causing large areas of Norfolk and Lincolnshire to be brought into cultivation.

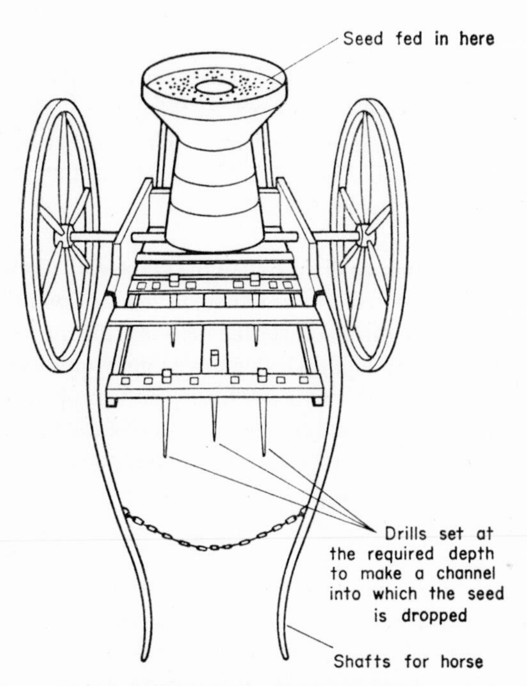

FIG. 7.2. Tull's horse-drawn drill.

In 1786 another important contribution was made to agriculture by Andrew Meikle who made a mechanical thresher. This device consisted of rollers through which the corn could be fed and which separated the seed from the straw and husks. For centuries the method of doing this essential job had been by flailing and winnowing in the air, a laborious, slow, and uneconomical system.

The eighteenth century also saw the beginning of scientific stock selection and breeding. Robert Bakewell (1725–95) applied deliberate selection for better strains by picking parents of outstanding quality and breeding them together rather than allowing the random mating within a herd or flock normally practised. He was successful in producing new and much improved strains of sheep, horses, and cattle, and during his lifetime the average weight of carcass sold at Smithfield (the main London market) was doubled largely by use of these better breeds (see also p. 108).

English farming had now reached a higher standard and efficiency than that of surrounding nations, and despite a rapidly expanding population we were able to sell much food abroad and became known as the "granery" of Europe. Indeed, the new prosperity of the farmers did much to finance the wars against Napoleon and allow them to be prosecuted to a successful conclusion.

Improvements were not, however, brought about without changes and upheavals. The enclosure of estates which had been going on from Tudor times began to rapidly increase and to apply to the common lands as well. There was so much inefficiency in the medieval system of open fields, and great stretches of common pasture and the new methods of farming worked much better where the land was enclosed. In the latter half of the eighteenth century the price of cereals and other crops was very high and it was as much to the advantage of the smallholder to enclose his land and to farm it efficiently as it was to the great estate owners. By 1750 more than half the 9000-odd parishes of the country had been enclosed, and by 1820 less than 3 per cent remained on the old system.

It is often said that enclosure drove many from the land and created cheap labour which made the industrial revolution possible. The fact is that population was rising steadily throughout this time in both town and country, and that relationship between agricultural and industrial revolution is very complex. It may well be true that it was the new prosperity of the country's agriculture, due amongst other things to the enclosures, that supported the non-farming population and, indeed, made the industrial revolution feasible. The real sufferings of the rural population were to come at the end of the Napoleonic wars when the price of corn fell disastrously.

The Development of Modern Farming

Agricultural Chemistry

While breeds were being improved and new machines invented, a further and more subtle change was at work to increase the productivity of the land. This change was due to the posing and solving of fundamental questions about the actual food requirements of the green plant. Of course there had been ideas from the very earliest times that new growth came from decay, and manures of various sorts had been used to dress the soil (see p. 99). Such application was more empirical than scientific, however, until the actual needs of the plant were known and could be specifically catered for with artificial fertilizers (i.e. chemicals rather than organic manures).

An attempt to find out the food of plants was made as early ago as 1600 by van Helmont who planted a young tree in a sealed and weighed box of earth and supplied it only with water for a period of 5 years. At the end of this time he weighed both the soil and the tree and found that the former had lost only a few ounces. This led him to conclude that the plant fashioned itself from water and the soil was of little importance except for support—a good example of the misinterpretation of scientific data and the making of a conclusion on insufficient evidence.

Some time after this Woodward investigated the growth rate of a water plant in various sorts of solution. He found that the least growth was made in rain water and the most in stream water that had been shaken up with soil. It was clear to him that the soil did contribute to the growth of plants, and he suggested the water acted only as a solvent.

Jethro Tull, inventor of the drill already described, thought that the roots of plants took in minute particles of soil, hence the latter should be rendered to as fine a tilth as possible. The idea of roots as a sort of extraverted digestive system equivalent to the intestine of mammals was quite common at this period.

A milestone in botanical physiology was Stephen Hales's *Vegetable Statics* published in 1727 in which were described many ingenious and accurate measurements of the ascent of the sap in trees and its loss to the atmosphere. Although his work did not bear directly on plant nutrition, he suggested that "in some way the air is turned into the substance of plants" and, indeed, went as far as to hazard that light might be of assistance in the process.

Priestley, the English chemist, showed that plants gave off oxygen and his work was furthered by Ingenhaiz who did numbers of experiments of photosynthesis which he published in 1778. He was the first to show that the plant took up carbon dioxide in the light.

The food of the plant also intrigued the brilliant French chemist de Saussure who was able, by experiment, to show that plants took up carbon dioxide and released oxygen by day but that the process was reversed at night. These results were obtained by shutting the living leaves of a plant within a flask and after a time had elapsed analysing the changes in composition of the air that had taken place. Beside these important discoveries de Saussure also found that the element nitrogen, a critical constituent of all living cells, was not taken up from the air but only from solution in the soil. In the early nineteenth century another Frenchman, Boussingault, followed up this work with the detailed analysis of plant ashes of various sorts of soil and manures and showed the elements present, thus establishing a practical composition for fertilizers.

Eighteen-forty saw the publication of *Organic Chemistry and its Application to Agriculture and Physiology* by the great German chemist Justus von Liebig. Among other things Liebig postulated that the growth of a plant would be controlled by an element that was present in limiting quantities and that in field conditions the growth of crops was in fact limited by the slow release of essential minerals from the soil.

Studying at this time under Liebig was the young Joseph Gilbert, and on his return to England in 1843 he teamed up with a well-to-do landowner in Hertfordshire who had an amateur interest in agricultural chemistry. This man, John Lawes, founded the first institute for research into plant growth, and here, at Rothamsted, he and Gilbert, together with an ever-increasing staff, worked on problems of plant yield and nutrition between 1843 and 1900. An early result of these experiments was the discovery that sulphuric acid applied to bone or to other sources of phosphate produced a soluble fertilizer of great value to plant growth. This substance was termed "superphosphate" and a patent taken out on its manufacture by Lawes. Although the results of the work at Rothamsted were more or less in agreement with the continental chemists, the English workers considered that von Liebig had underestimated the importance of an adequate nitrogen supply for the plant. Much of the work they did was concerned with the sources and rate of uptake of nitrogen from the soil. On plots of land isolated and studied over many seasons they found soil without manuring gave only some 13 bushels of cereal crop each year per acre while with a complete dressing of fertilizer including adequate supplies of nitrogen some 32 bushels could be raised.

Boussingault had shown that leguminous crops actually added to the nitrogen content of the soil, and this was confirmed by experiments at Rothamsted although it was not until the end of the nineteenth century that the means whereby they did this were discovered. It was found that the small nodules on the roots of peas, clovers, and other legumes contained bacteria that could actually fix the nitrogen from the atmosphere and convert it into a form which the plant could use.

The various discoveries of these pioneer agricultural chemists led to immediate application in western Europe, and sources of artificial fertilizers began to be opened up and exploited in quantity. Sodium nitrate was imported from Chile, where large natural deposits occur, and sources of naturally occurring phosphate were utilized in other countries. Ammonium nitrate, a by-product of the gas works, began to find commercial application as did the basic slag which remained as waste from the Bessemer process of steel manufacture. In the present century atmospheric nitrogen and oxygen are combined together across an electric arc, or by catalysis, to yield synthetic nitrogenous fertilizers, and many millions of tons are produced annually in this way.

The twentieth century has also seen a more sophisticated use of the different chemicals available, and specific mixtures are applied to crops and soil according to their nature and requirements. The original chemical fertilizers were often corrosive or unpleasant to handle, and it is now customary to make use of clean granulated products with specially balanced constituents.

For us to understand just how important the science of crop fertilizers has become we may consider a few figures. Under the four-course rotation of crops of the seventeenth century approximately 60 lb of nitrogen were taken up from the soil per acre per year; today an average crop would be expected to remove nearly eight times this amount and the yields would be proportionate. Today one farmer in Britain grows sufficient to support 12–15 other men while in the seventeenth century nine farmers could support only themselves and one other man. In 1900 only 19,000 tons of nitrogenous fertilizer was used in the whole country but by 1956 this had risen to 291,000 tons and it is estimated that this quantity could profitably be doubled.

Beside the "big six" minerals, i.e. nitrate, sulphate, phosphate, calcium, potassium, and magnesium, the agricultural chemists have found a whole lot of other elements, sometimes only needed in minute amounts but whose presence is necessary for healthy plant growth. These latter are called trace elements and include manganese, zinc, boron, etc. If they are not present the plant will show symptoms of deficiency diseases and will not grow properly or set seed. With modern knowledge such diseases can be diagnosed and remedied by the addition of the appropriate element, and this technique is used not only on a small scale but, as in Australia and New Zealand, over great areas of land.

A further progress in fertilizers was made very recently when it was discovered that certain organic compounds, called chelates, had the property of taking up minerals that would normally be locked insolubly in the soil and making them available for plant roots. Such chelates have found successful application in the United States.

Mechanization

Apart from Tull's drill and harrow and Meikle's threshing machine already described, there had been very few successful new devices brought into general use by the end of the eighteenth century. The basic techniques of preparing the soil, reaping, and threshing were still done in much the same way as in the Middle Ages.

The period 1800 up to and including our own times has seen the invention of great numbers of machines which have revolutionized almost every aspect of practical farming, and have done much to contribute to the high yields and efficiency of modern agriculture.

In the closing years of the eighteenth century Eli Whitney of the United States made a cotton gin which was able to separate the fibres from the seed and the invention did much to help the great cotton growing southern states. This gin was interesting in that interchangeable parts were used—a characteristic of our modern mass production techniques. Another important invention, again from the States, was a workable reaping machine which replaced the slow and laborious use of the scythe in cutting cereal crops. There had, in fact, been many sorts of reaper made; indeed Pliny (A.D. 23–79) records the Gauls as using a rotary cutting implement, but there seems no doubt that the first really effective one was that introduced by McCormick in 1831. Up to this time a reaping team, consisting of a scytheman and four collectors and binders, could cut some 2 acres a day if they worked fast. The mechanical reaper allowed one man and a horse to cut some 6 acres in the same time.

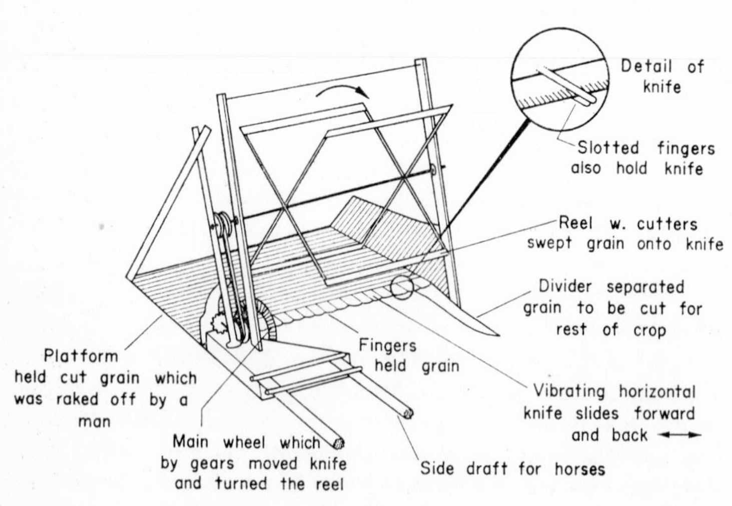

FIG. 7.3. McCormick's reaper.

The principles of McCormick's invention are shown in Fig. 7.3. A vibrating horizontal knife, running between slotted fingers, cut the corn, and a rotating reel and cutters pushed the severed portion back onto a platform whence it could be removed by a long rake. The power for the knife and reel came from the single large wheel on which the implement moved being drawn along by a horse. Later a seat was built on so that the driver could be carried along on the reaper, and by 1845 mechanical binders were also attached.

Trevithick had made his steam-engine in 1812, and by the middle years of the century pairs of steam-engines moving down to the edges of fields were being used to pull farm instruments across the ground by long warps. The method had very obvious limitations, and a few years later movable traction-engines came in and began to replace the horse as a source of power. Steam-engines are more powerful than horses, hence it was possible to make larger ploughs and cultivators.

In the closing decades of the last century inventions came thick and fast. To the reaper there was added a device for winnowing, i.e. separating the grain from the chaff, and by 1880 reaper, thresher, winnower, and binder have all been incorporated into one great machine, driven with its own power sourçe—the combine harvester. The internal combustion engine was actually invented in the late nineteenth century, and the first oil-fueled tractors were made in the United States in 1889. The widespread use of tractors came in, in Europe, during the First World War (1914–18).

Since the First World War there has been the refinement of machines, the use of new sources of power such as electricity, and the mechanization of almost every conceivable process that might be expected to take place on a farm.

Tractors of various sizes and powers and with hydraulic control of towed gear now pull ploughs whose depth and angle of attack can be adjusted to a particular soil. Rotary cultivators and bull-dozers clear and reclaim land into which drainage pipes are mechanically inserted. Harrows and rollers remain the same, but there are now drills for planting all sorts of crops and inserting fertilizer at one and the same time. The combine harvester deals with and processes cereals, while other machines gather in crops that vary from potatoes to peas. Not only fertilizers but hormone weed-killers and pesticides can be sprayed from ground machinery as well as from helicopters, while livestock can be fed, and even—in the case of cows—milked mechanically. We have certainly come a long way from the time of Jethro Tull and Andrew Meikle.

New Livestock and Crops

Under the medieval system of common land and grazing rights there was little opportunity for improving stock. As it was said any animal could well be "a haphazard union of nobody's son with anybody's daughter" and it was not possible to maintain any sort of line or purity of breed. With the coming of enclosures and winter feeds there was a growing interest in improving livestock from the half-starved and unhealthy animals that had dotted the English landscape for so many hundreds of years.

When Robert Bakewell started his breeding work in the latter part of the eighteenth century there were nearly as many breeds and varieties of stock as there were counties. Bakewell set as his aim certain desirable characters, such as good thickness at the joints, efficient food conversion, and early maturity, and he selected and bred towards these ends. Good was crossed with good, and skeletons and joints were preserved so comparisons could be made between off-spring and parent. After perfecting a male he would hire it out rather than sell it, and keep a record of its progeny. Sometimes his experiments were unsuccessful; thus in breeding for better meat in Lancashire longhorns he produced a cow that was almost milkless.

The methods of selection for pure lines are still used, and in his own time he made a very substantial contribution to livestock yield (see p. 101).

It is common nowadays to have "stock societies" where pedigrees and records are kept and which encourage development of a particular breed. Champion stock are exhibited and exported from one country to another, and by the techniques of artificial insemination one male can father very many progeny. A recent estimation suggests that some 6,000,000 cattle are produced annually in the United States by this method.

Plant breeds are also continually being improved by selection and special breeding. The notion that plants had a form of sexuality appears about 1694, and a few years later the first plant hybrid was made using a carnation and a sweet william as parents. In the period between 1759–1835 a number of new fruits were made by deliberate crosses.

An obvious milestone in plant and animal breeding was the rediscovery of Mendel's laws of inheritance (*c.* 1900, see p. 47), and an early application was made by Roland Biffen at Cambridge. This example also serves to illustrate the sort of ways in which immunity to disease can be bred into a crop.

Biffen took the rust-resistant but low-yielding wheat called Ghurka and crossed it with Squarehead's Master, a good yielding variety susceptible to rust. The resulting plants were all susceptible and of low yield, these being the dominant characters. On interbreeding this generation, just as Mendel had done with his pea plants, ratios of the two characters appeared 9, 3, 3, 1, the 1 plant in 16 showing both resistance and good yield. This variety was called Little Joss and was widely used for some years.

Using the symbols and layout explained in Chapter 4 we can represent the breeding of Little Joss as follows:

Let R be rust susceptible and r rust resistant (i.e. R is dominant), Y a poor yielding character, and y a good yielding one (Y dominant);

Parents	$rrYY$		×	$RRyy$
	rY	rY	Ry	Ry
F_1	$rYRy$		$rYRy$	
	RY	Ry	rY	ry
RY	$RYRY$	$RYRy$	$RYry$	$RYry$
Ry	$RyRY$	$RyRy$	$Ryry$	$Ryry$
rY	$rYRY$	$rYRy$	$rYrY$	$rYry$
ry	$ryRY$	$ryRy$	$ryrY$	$ryry$

Result: 9 rust susceptible poor yield. 3 rust susceptible good yield. 3 rust resistant poor yield. 1 rust resistant good yield $ryry$ (variety Little Joss).

Charles Darwin himself had been interested in the effects of inbreeding and selection as a means of bringing about biological change in a species and amongst other plants he had worked on corn (maize). One important finding was that inbreeding led to loss of vigour in the offspring as compared with outbreeding, and this idea was investigated further in the United States in the first decade of this century.

Outstanding were the names of East and Shull who independently discovered that although inbred corn did lose vigour, when two inbred pure lines were hybridized the resulting offspring were of high quality. This sort of "single-line hybrid", although giving fine yields, was not easy to exploit commercially, and a further advance was made about the time of the First World War, when two hybrids were successfully crossed, making a double-line cross (Fig. 7.4). The yields from these double hybridizations were a great improvement on any previous varieties, and the principles employed have now been widely used in the development of new crops.

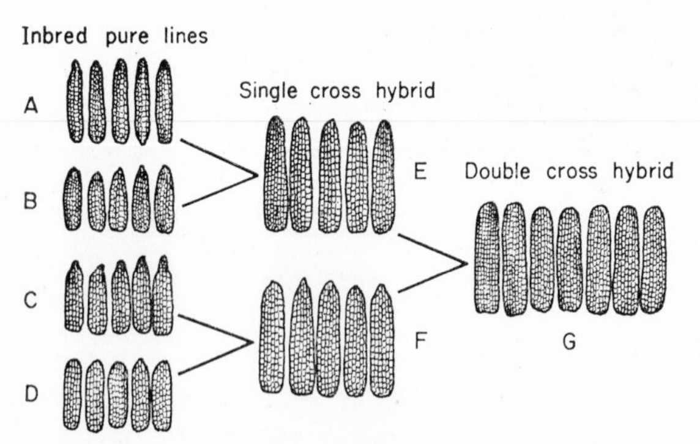

FIG. 7.4. Hybrid maize.

If we look at a table of breeding programmes prepared from the current work of just one plant breeding station (Cambridge Plant Breeding Institute) and, of course, there are now many of these throughout the world, some idea of the scope and aims of the modern crop improvers can be realized.

Crop	**Work in hand, 1965**
Cereal	
Wheat	Spring and winter strains are being bred which give high yield with hard milling grain for flour. Resistance to powdery mildew, yellow rust, and eyespot as well as other less important diseases is incorporated in these strains.
Barley	Varieties with special hardiness for autumn sowing are being investigated together with improvement of malting quality. Resistance to mildew, brown and yellow rust, and scald are important features of the new strains.
Oats	High yield with low husk content together with resistance to powdery mildew and oat-stem eel-worm are the features towards which current breeding programmes are designed.

(Selected lines of cereals which have proved satisfactory in trials are sent on to the National Institute of Agricultural Botany and if, after further rigorous trials, they prove satisfactory, put on the national list recommended to farmers.)

Sugar-beet	Important features undergoing investigation are resistance to bolting in early-sown beet. New varieties which can stand infection with the yellow virus that normally attacks beet have been developed. Selection for high sugar content is also practised.
Potatoes	The aims are to produce varieties, especially of first earlies, with a better quality and yield than Arran Pilot, Majestic, and King Edwards at present grown in Great Britain. Incorporated into these varieties should be resistance to the root eel-worm and to wart and scab diseases. Successful eel-worm resistant strains (in particular Maris Piper) are being multiplied for release to the public.
Kales	Double and triple hybrid crosses (see above) have been made to increase vigour of strains. A short-stem hybrid suitable for strip grazing is being multiplied for general release. In selection attention is paid to leaf retention, winter-hardiness, and resistance to disease and aphid attack.
Lucerne	Production of new strains has been disappointing and an attempt at a double-cross hybrid is being made. Resistance to wilt is found in wild species from North Africa and the Caucasus is being incorporated into present strains used in this country.
Clover	Addition of resistance to stem eel-worm, mildew, and scorch is being studied.

Grasses	Strains of rye grass, cocksfoot, etc., are being bred with extra ability to stand drought and with increased winter hardiness.

Besides these problems fundamental work on chromosome transfer is on hand.

The Control of Disease

Any continuous large-scale production of plant or animal life in a closed and crowded community is to a large extent against the way in which living things are distributed in nature. In such artificial communities the parasites and diseases of the specific organism being reared are also likely to flourish.

As we have already seen in Chapter 4, the notion that diseases were due to evil fogs, devils, and the like had given way by the latter half of the nineteenth century to the germ theory developed by Pasteur and his fellow bacteriologists. While the applications of this knowledge were done more immediately to animals and to man himself, it was also of great importance in understanding the nature of many plant diseases.

The main causative agents of plant diseases are the fungi, a group of plant organisms that do not photosynthesize but who live either as saprophytes on decaying organic matter or as parasites on living tissues. The classification and morphology of these had been worked on some time before they were recognized as being associated with disease, and even when fungal strands and fruiting bodies were found in the lesions of sick plants, or growing out over their surface, their presence was thought to be some manifestation or symptom of disease rather than its cause. As early as 1807 Prevost had suggested that the bunt disease of wheat was due to a fungus, but his views were rejected.

Eighteen-fortyfive saw the start of the Irish potato famine caused by the dying off of the haulm and tuber and loss of the staple crop upon which the peasants largely depended. A commission set up to investigate the cause of the disease arrived at the conclusion that the cool moist weather was responsible and the appearance of fungal growths on the dying plants was a secondary effect. This finding was questioned by the mycologist Berkley (1846), who put forward the idea that it was the fungus, then called *Botrytis infestans*, which actually caused the disease. A few years later Anton de Bary published a comprehensive paper on fungi including a section on those that can cause disease such as smuts and rusts on cereals. In a series of brilliant papers de Bary confirmed Berkley's ideas about the potato blight putting the causative agent into the newly created genus *Phytophthora*. By finding the life cycles and intermediate hosts and the conditions under which these various harmful fungi thrived, de Bary was able to point to the ways in which they could be controlled.

One of the most famous, and still widely used, means of combating fungal diseases is by use of Bordeaux mixture. This mixture of copper sulphate and lime is very toxic to fungi and was first introduced by Millardet, a one-time pupil of de Bary's, in 1882. This fungicide does cause some damage to leaves and fruits, however. From the successful use of sulphur washes, lead arsenates, and mercuric chlorides developed in the early part of this century to the very modern organic chemicals, such as chloranil, the farmer now has a wide range of specific fungicides that he can apply.

A further causative agent of plant disease came to light in 1892 when Iwanowsky (see p. 90) found that the sap from mosaic-infested tobacco contained an ultramicroscopic and filterable infective organism, later to be termed a virus. Since that time many virus diseases have been discovered.

Summary

In the short account of agriculture above we have considered how this fundamental human skill has progressed from its early beginnings to the present time.

At first man the hunter, or the nomad, was dependent on the natural productivity of an uncontrolled environment. Whether the individual or the tribe lived or perished was largely a matter of chance; there was no certain or regular supply of food.

The basis of the great civilizations of the Egyptians, the Romans, and others was that agriculture had become established at least to the point of self-sufficiency. The majority of men worked on the land supporting themselves with a little over. On this latter depended the wealth and structure of their society.

This pattern of bare self-sufficiency continued through the Middle Ages until the great agricultural revolutions brought about by social and technological change. After the fourteenth century, at least in Britain, agriculture becomes the provider, indeed more and more so. The wealth of the nation and its capacity to become industrialized and to emerge into modern times was based on the prosperity of over-production in agriculture.

Today the pattern has changed again. The scientific revolution of the last century has had its own widespread effects on agriculture which, in the decades since the war, has become increasingly automated and efficient. Very high yields of all sorts of crops can be achieved by modern methods, and such yields are found in many parts of the world. Yet the human race increases at a rate which means that much of it is undernourished or actually starving, or is liable to starve should a single growing season be poor. In certain parts of the earth the standard farming practices and the yields they give are still reminiscent of medieval Christendom, and the applications of scientific knowledge are not yet applied.

Questions on Chapter 7

1. How has the discovery of the nitrogen cycle led to increased productivity from the soil?
2. Describe the basic steps in sowing and harvesting a cereal crop. In what way do our modern methods differ from those used in the Middle Ages?
3. How were the discoveries of Mendel made use of in plant breeding?

Bibliography

A List of Inexpensive and Readable Books that are Suitable for Further Study

General Reading on the History of Science and Scientific Method

The Major Achievements of Science. McKENZIE. Cambridge.
The Origins and Growth of Biology. ROOK. Pelican.
A Short History of Science. DAMPIER. Cambridge.
Reason and Chance in Scientific Discovery. TATON. Scientific Book Guild.

Chapter 2. Classification

There is little on this that is likely to be of interest to the non-specialist although there are worthwhile extracts from Linnaeus's work in *The Origins and Growth of Biology*. For the more specialized reader *Animal Diversity* by HANSON, published by Prentice-Hall, and *New Concepts in Flowering Plant Taxonomy* by HESLOP-HARRISON, may be of use. The difficulties of constructing a sure animal classification are also excellently described in Kerkut's *Implications of Evolution*.

Chapter 3. Evolution

The Darwin Reader. BATES and HUMPHREY. Macmillan.
Darwin's Century. EISELEY. Scientific Book Guild.
(These are much easier going than Darwin's *Origin of Species*.)
The Ideas of Biology. BONNER. Eyre & Spottiswoode.
Evolution and its Implication. KELLY. Scientific Book Club.
The Mechanism of Evolution. DOWDESWELL. Heinemann.
Evolution. British Museum (Natural History).
A Hundred Years of Evolution. CARTER. Sidgwick & Jackson.

Chapter 4. Heredity

Elementary Genetics. GEORGE. Macmillan.
Genetics. KALMUS. Pelican.
Genetics for Medical Students. FORD. Methuen.

Chapter 5. *Anatomy, Physiology, and Surgery*

British Medicine. McNAIR WILSON. Collins.
A Short History of Medicine. POYNTER and KEEL. Scientific Book Club.
The Conquest of Pain, Anaesthesia. BANKOFF. MacDonald.
The Early History of Surgery. BISHOP. Scientific Book Guild.
William Harvey. CHAUVOIS. Scientific Book Guild.
Social Biology. DALE. Heinemann.

Chapter 6. *The Germ Theory of Disease*

The Microbe Hunters. DE KRUIF. Harcourt, Brace & Co.
Louis Pasteur. NICOLLE. Scientific Book Guild.
The Life of Sir Alexander Fleming. MAUROIS. Cape.
Beyond the Microscope. SMITH. Pelican.

Chapter 7. *Agriculture*

Great Men of Modern Agriculture. CANNON. Macmillan.
Short History of Farming. WHITLOCK. Baker.
A History of Agriculture. FRANKLIN. Bell.
Vegetable Statics. HALES. Scientific Book Guild.
The Land. HIGGS. Studio Vista.

List of Persons and their Biological Work

Addison, T. (1793–1860): disease associated with adrenals.

Aristotle (384–322 b.c.): biological classification and philosophy.

Avery, O. *et al.*: (20th cent.) isolation of genetic material from bacteria.

Bacon, R. (1214–94): philosophy of science.

Bakewell, R. (1725–95): improvement of stock.

Banting, F. G. (1891–1941): discovery of insulin.

Bary, H. A. de (1831–88): life histories of pathogenic fungi.

Bateson, W. (1861–1926): the nature of genetic variation.

Beadle, G. W. (20th cent.): connection between genes and enzymes.

Behring, E. A. von (1854–1917): immunization against diphtheria.

Beneden, E. Van (1845–1910): discovery of meiosis.

Bentham, George (1800–84): plant classification.

Berkley, G. (1800–71): work on life cycles of pathogenic fungi.

Bernard, Claude (1813–78): physiologist.

Best, C. H. (1899–): co-worker with Banting on insulin.

Boussingault, J. B. J. D. (1802–87): agricultural chemistry.

Boyle, R. (1627–91): composition of respiratory gases.

Bridges, C. (1889–1938): linkage of genes on chromosomes.

Bruce, D. (1855–1931): work on sleeping sickness and its vector.

Brunfels, O. (1489–1534): description of plant species.

Buffon, G. L. L. (1707–88): naturalist and philosopher of biology.

Candolle, A. P. de (1778–1841): classification.

Celsus (1st cent. a.d.): medicine and surgery.

Cesalpini, A. (1519–1603): plant classification.

Cook, J. (1728–1779): use of lime juice to prevent scurvy.

Crick, F. (20th cent.): the code of the gene.

Cuvier, G. L. C. (1769–1832): classification and palaeontology.

Darwin, C. (1809–82): the concept of evolution by natural selection.

Darwin, E. (1731–1802): philosopher of biology.

Davy, H. (1778–1829): anaesthetic gases.

Descartes, R. (1596–1650): philosopher of science.

Dioscorides, P. (1st cent. a.d.): herbalist.

DOBZHANSKY, T. (20th cent.): mathematics of natural selection.
DOMAGK, G. (1895–): discovery of sulphonamides.
EAST, E. (1879–1938): plant hybridization.
EHRENBERG, C. G. (1795–1876): description of protozoa.
EHRLICH, PAUL (1854–1915): chemotherapy.
EMPEDOCLES (3rd cent. B.C.): doctrine of humours.
ERASISTRATUS (3rd cent. B.C.): anatomy and physiology.
FABRICUS, H. (1537–1619): worked on blood system; tutor to Harvey.
FARADAY, M. (1791–1867): anaesthetic gases.
FISHER. R. (20th cent.): mathematics of selection in populations.
FUNK, C. (1884–): isolation of vitamins.
GALEN (A.D. 130–200): anatomy, physiology, and medicine.
GALTON, F. (1822–1911): studies on heredity.
GESNER, K. von (1516–65): natural historian.
GILBERT, J. M. (1817–1901): co-founder of Rothamsted Experimental Station.
GRAAF, R. de (1641–73): description of the mammalian ovary.
GRASSI, G. B. (1854–1925): work on malaria.
HAECKEL, H. (1834–1919): embryology and evolution.
HALES, S. (1677–1761): plant and animal physiology.
HALLER, A. von (1708–77): human physiology and anatomy.
HARVEY, W. (1578–1657): the circulation of the blood.
HELMHOLTZ, H. L. F. von (1821–94): invention of the ophthalmoscope.
HELMONT, J. B. Van (1577–1644): plant nutrition experiments.
HEROPHILUS (3rd cent. B.C.): medicine.
HIPPOCRATES (4th cent. B.C.): medicine.
HOOKE, R. (1635–1703): description of cells.
HOOKER, J. D. (1817–1911): botanical classification.
HOPKINS, F. G. (20th cent.): studies on nutrition.
HUNTER, J. (1728–93): medicine.
HUTTON, J. (1726–97): geology.
HUXLEY, T. H. (1825–95): zoology and evolution.
INGENHOUSZ, J. (1730–99): the photosynthesis of plants.
JENNER, E. (1749–1823): smallpox vaccination.
KETTLEWELL, H. (20th cent.): melanism in moths.
KOCH, R. (1843–1910): bacteriologist.
LAMARCK, J. B. P. A. de M. (1744–1829): classification and evolution.
LANDSTEINER, K. (1868–1943): blood groups.
LAVERAN, C. L. A. (1845–1922): malarial parasite.
LAWES, JOHN (1814–1900): founder of Rothamsted Experimental Station.
LEDERBERG, J. (20th cent.): isolation of nucleic acid.
LEEUWENHOEK, A. Van (1632–1723): microscopist.
LIEBIG, J. (1802–73): organic and agricultural chemist.
LIND, J. (1716–94): nutrition.

LINNAEUS, C. (1707–78): classification.
LISTER, J. (1827–1912): antiseptic surgery.
LYELL, C. (1797–1875): geology.
McCORMICK, C. (1809–84): invention of the reaper.
MAGENDIE, F. (1783–1855): the nervous reflex arc.
MALPIGHI, M. (1628–94): microscopist.
MALTHUS, E. T. (1766–1834): essay on population.
MANSON, P. (late 19th cent.): malaria.
MEIKLE, A. (*c.* 1750–1800): threshing machine.
MENDEL, G. (1822–84): genetics.
METSCHNIKOFF, E. (1845–1916): the discovery of phagocytes.
MORGAGNI, G. B. (17th cent.): diseases and their classification.
MORGAN, T. H. (1866–1945): genetics of *Drosophila*.
MÜLLER, H. J. (1890–): gene mutations.
PASTEUR, L. (1822–95): bacteriology and disease immunization.
PAVLOV, I. (1849–1936): function of the nervous system.
PLINY (A.D. 23–79): agricultural writings.
PRIESTLEY, J. (1733–1804): composition of respiratory gases.
RAY, J. (1628–1705): plant classification.
REDI, F. (1626–97): spontaneous origin of life disproved.
REED, W. (1851–1902): transmission of yellow fever.
ROSS, R. (1857–1932): transmission of malaria.
RÖNTGEN, W. K. (1845–1925): discovery of X-rays.
SAUSSURE, T. de (1767–1845): plant physiology.
SCHLEIDEN. M. (1804–81): cell theory.
SCHWANN, T. (1810–82): cell theory.
SERVETUS, M. (1511–53): anatomy and physiology.
SHERRINGTON, C. S. (1861–1952): physiology of the nervous system.
SHULL, G. (1874–1954): plant hybridization.
SIMPSON, J. Y. (1811–1870): anaesthetics.
SMITH, T. (20th cent.): disease transmission.
SMITH, W. (1769–1839): geology.
SPALLANZANI, L. (1729–99): work on the spontaneous germination of life.
STURTEVANT, A. (20th cent.): mapping of chromosomes.
SUTTON, W. (1876–1916): genetics.
TATUM. E. (20th cent.): genes and enzymes.
THEOPHRASTUS (*c.* 380–287 B.C.): plant classification and description.
TOWNSHEND, C. (1674–1738): agricultural practice.
TULL, J. (1674–1741): invention of the seed drill.
VERSALIUS, A. (1514–64): anatomy and physiology.
VINCI, L. da (1452–1519): anatomy and physiology.
VRIES, H. de (1848–1935): genetics.
WALLACE, A. (1823–1913): zoology and evolution.
WATSON, J. B. (1878–): chemistry of nucleic acid.
WEISMANN, A. (1834–1914): continuity of the germ-plasm.
WHITNEY, E. (1761–1825): invention of the cotton gin.
WILLUGHBY, F. (1635–72): animal classification.

Glossary of Biological, Scientific, and Agricultural Terms

ammonite: extinct cephalopod molluscs of much use in the dating of rocks.

amphilia: a class of the vertebrate or back-boned animals which have legs and reproduce in water and includes newts, frogs, and toads.

amputation: the removal of a limb or part of a limb by surgery.

analogy: organs which share a common function (e.g. wings of insects and birds) as distinct from homologous organs which have a common origin (e.g. wing of bird and front limb of man.)

anatomy: the study of gross structures in the body.

annelid: a phylum of worms whose bodies are clearly segmented—includes the earthworm.

aseptic: where no micro-organisms exist, usually refers to modern surgery which is carried out in such conditions and contrasts with antiseptic surgery where chemicals are applied during operations to kill bacteria.

antibiotic: a chemical prepared from a living plant, usually a soil fungus such as penicillium, which destroys bacteria.

antiseptic: the application of chemicals such as carbolic acid for the destruction of micro-organisms (cf. *aseptic*).

bacteriophage: a micro-organism which parasites bacteria and which is intermediate in size between bacteria and viruses.

bacteria: microscopic organisms which are present all over the earth. They are mostly harmless and bring about decay of organic matter, but some, such as the cholera and typhoid bacteria cause diseases in man and other animals.

binomial classification: the naming of an organism with two names, a generic name and a specific name. Together these two names will be unique for that organism—the system was made popular by Linnaeus in the eighteenth century.

bivalve: a type of mollusc with two valves, e.g. a cockle.

carbolic acid: a phenolic organic substance that has strong germicidal properties.

cephalopods: molluscs related to and including the octopus and squid.

chromosome: thread like parts of the nucleus which contains the genes or character determining molecules. The number of chromosomes is fixed for a species and they are normally composed of desoxyribose nucleic acid (DNA).

class: a large natural grouping of animals or plants. It is itself a sub-division of a phylum and includes within it orders, families, and genera.
combine harvester: machines, now often self-powered, which cut, thresh and bag cereal crop in one operation.
crustaceans: members of the jointed limbed animals or arthropods, related to insects but with a massive shell and at least five pairs of legs.
deficiency disease: a disorder due to the lack of some essential factor such as a vitamin, first class protein, or mineral salt from the diet.
desoxyribose nucleic acid (DNA): the chemical from which the genes and chromosomes are made. It has a structure which allows it to code large amounts of information which are used in the activities of the cell.
dicotyledons: those flowering plants with two embryonic leaves in their seeds, e.g. buttercup, bean.
dominant character: where contrasting characters or genes are found the one that shows itself when in the heterozygous condition is said to be dominant.
echinoderms: a group of animals that are radially symmetrical, marine and covered with a shelly skin. They include starfishes and sea urchin.
ecologist: one who studies the relationship of animals or plants to their natural habitats.
endocrines: glands which secrete a chemical, called a hormone, into the blood whence it travels to the part of the body it affects. Thus the pituitary secretes pituitrin and the thyroid thyroxin.
evolution: the concept that life has changed over vast periods of time from some simpler state to the present day complexity of forms.
fertilizer: a substance which increases the productivity of soil (or water); can be natural, such as manure, or artificial such as superphosphates.
filial generation: in breeding experiments the first generation offspring are called the first filial or F_1 and the second generation the second filial or F_2.
fossil: the cast or remains of a living organism that has become preserved.
fungi: a group of plants who do not photosynthesise and which are normally saprophytic feeding on dead organic matter. Some, such as potato blight, are important parasites of higher plants.
gene: a heredity factor borne on a chromosome.
genetics: the study of inheritance.
genus: a group of very similar animals or plants.
germ: a rather loose term meaning harmful micro-organism.
germplasm: the part of the animal or plant body not concerned with sexual reproduction.
harrow: an instrument for breaking up soil into small particles for cultivation.
heredity factor: a character which is transmitted from one generation to another (cf. *gene*).
homology: organs which share a common origin and which are not necessarily analogous (see *analogy*).

homunculus: a tiny human being thought to be present in egg or sperm.

hormone: a chemical secreted into the blood by an endocrine gland. Thus insulin is a hormone from the pancreas.

hybrid: the offspring produced by crossing parents with disimiliar characters usually of different species.

immunization: becoming resistant to a disease either by activity of ones own body defences or passively by inoculation of antibody from another person or animal.

infusona: microscopic life, both animal and plant, that appears in cultures that have been standing for some time.

inheritance of acquired characters: the belief that characters gained during the life time of an organism can be passed on to its descendants.

insect: animals with external skeletons and jointed limbs whose bodies are divided into three parts, head, thorax, and abdomen. They have wings and three pairs of legs.

interferon: substance made in living cells against virus infections.

invertebrate: an animal without a backbone.

linkage: where two or more genes are situated on the same chromosome they are said to be linked. The number, or group of linked genes thus corresponds to the number of chromosomes.

marsupial: primitive mammals, largely confined to Australasia, that rear their young in a pouch, and do not have a placenta, e.g. kangaroo.

meiosis: the division of genetic material that takes place at the formation of sex cells (or spores in some plants). It results in a halving of the material and a mixing up of the parental genes.

Mendel's laws: the laws of definite proportions and independent segregation of characters in the second filial generation which laid the basis for a quantitative study of heredity.

metabolism: the sum of all the physiological activity of a living organism.

micro-biology: the study of microscopic organisms such as bacteria.

micro-organism: any organism that is small enough to necessitate use of a microscope for its examination.

microscope: an instrument for magnifying tiny objects. The light microscope can magnify up to some 2000 times, but for greater enlargements an electron microscope (one that uses a beam of electrons and electromagnets as lenses) must be used.

molluscs: animals usually with shells and a large muscular foot, e.g. snails, cockles.

monocotyledon: a major subdivision of the family of flowering plants which have only one embryo leaf in the seed, e.g. grass.

monotreme: a primitive type of mammal which lays eggs—found in Australasia; e.g. duck-billed platypus.

mutation: a chemical change in a gene, induced by radiation or occurring naturally and which is inherited.

natural selection: the process of competition among living organisms whereby favourable varieties tend to survive and leave offspring and unfavourable perish. It is part of the mechanism of evolution.

neo-Darwinism: the Darwinian idea of natural selection combined with modern knowledge of gene mutation and selection. Neo-Darwinism is the present accepted mechanism for evolution.

ophthalmoscope: a diagnostic instrument for examination of the inside of the eye.

order: the subdivision of a class which itself consists of a number of related families. Thus ants, bees, and wasps are all of the order Hymenoptera, all having membrane like wings.

palaeontology: the study of fossils.

pangenesis: an incorrect view as to the nature of inheritance which supposed heredity material entered the sex cells from all parts of the body.

pasteurization: the heating of a substance up to *c.* 145°F which kills off most active bacteria. Used in the preparation of beer and treatment of milk.

penicillin: an antibiotic derived from the fungus *Penicillium.*

phagocytes: white blood corpuscles which can destroy germs.

physiology: the study of function.

phylogeny: ancestory.

Pliocene: a geological period of the Canozoic era—an important time in the evolution of mammals.

polyploid: having more than two sets of chromosomes.

prognosis: the foretelling of the course of an infection or disease.

protozoan: a single-celled animal such as amoeba.

recessive character: heredity character which can only show itself when homozygous, i.e. when no dominant contrasting character is present.

reflex arc: the way in which sensory and motor nerves are connected across the spinal cord so that stimulation of an appropriate sense organ causes an immediate response.

reptile: a class of the vertebrates which are characterized by laying eggs, having four legs and scaly skins, e.g. lizard.

respiration: the processes in the body of a plant or animal which release energy from food. Not the same as breathing which is a part of respiration.

seed drill: a machine for planting seeds in rows and at a required depth.

species: a group of organisms that naturally interbreed and produce fertile offspring.

soma: cells of the body other than sex cells.

spontaneous origin of life: the incorrect idea that life could start by itself from putrifying matter.

stethoscope: an instrument for listening to the heart or lungs.

suppuration: the going bad or the infection of a wound or a surgical incision.

systematist: one engaged on biological classification.

taxonomy: study of the relationship of one species with another and the making and correct naming of related groups.

therapeutic: healing measures applied to disease.

threshing: process of separation of the grain from the chaff or husk.

trilobites: ancient group of extinct arthropods, important zone fossils of the palaeozoic era.

vector: an agent which transfers disease from one organism to another. Thus the mosquito is a vector of malaria and yellow fever.

vitamin: essential food factor required in small quantities to maintain health.

Index